Broadband Terahertz Devices and Communication Technologies

Broadband Terahertz Devices and Communication Technologies

Editors

Lu Zhang
Xiaodan Pang
Prakash Pitchappa

MDPI • Basel • Beijing • Wuhan • Barcelona • Belgrade • Manchester • Tokyo • Cluj • Tianjin

Editors

Lu Zhang
College of Information
Science and Electronic
Engineering
Zhejiang University
Hangzhou
China

Xiaodan Pang
Department of Applied
Physics
KTH Royal Institute of
Technology
Stockholm
Sweden

Prakash Pitchappa
Institute of Microelectronics
Agency for Science,
Technology and Research
Singapore
Singapore

Editorial Office
MDPI
St. Alban-Anlage 66
4052 Basel, Switzerland

This is a reprint of articles from the Special Issue published online in the open access journal *Micromachines* (ISSN 2072-666X) (available at: www.mdpi.com/journal/micromachines/special_issues/terahertz_communication_technologies).

For citation purposes, cite each article independently as indicated on the article page online and as indicated below:

LastName, A.A.; LastName, B.B.; LastName, C.C. Article Title. *Journal Name* **Year**, *Volume Number*, Page Range.

ISBN 978-3-0365-7743-2 (Hbk)
ISBN 978-3-0365-7742-5 (PDF)

© 2023 by the authors. Articles in this book are Open Access and distributed under the Creative Commons Attribution (CC BY) license, which allows users to download, copy and build upon published articles, as long as the author and publisher are properly credited, which ensures maximum dissemination and a wider impact of our publications.

The book as a whole is distributed by MDPI under the terms and conditions of the Creative Commons license CC BY-NC-ND.

Contents

Lu Zhang, Xiaodan Pang and Prakash Pitchappa
Editorial for the Special Issue on Broadband Terahertz Devices and Communication Technologies
Reprinted from: *Micromachines* **2023**, *14*, 1044, doi:10.3390/mi14051044 1

Fengyuan Yang, Prakash Pitchappa and Nan Wang
Terahertz Reconfigurable Intelligent Surfaces (RISs) for 6G Communication Links
Reprinted from: *Micromachines* **2022**, *13*, 285, doi:10.3390/mi13020285 5

Yukang Feng, Han-yu Tsao and N. Scott Barker
THz MEMS Switch Design
Reprinted from: *Micromachines* **2022**, *13*, 745, doi:10.3390/mi13050745 29

Xu Wang, Ting-Peng Li, Shu-Xia Yan and Jian Wang
Room-Temperature CMOS Monolithic Resonant Triple-Band Terahertz Thermal Detector
Reprinted from: *Micromachines* **2023**, *14*, 627, doi:10.3390/mi14030627 43

Yuan Feng, Xingwang Bian, Bowen Song, Ying Li, Pan Pan and Jinjun Feng
A G-Band Broadband Continuous Wave Traveling Wave Tube for Wireless Communications
Reprinted from: *Micromachines* **2022**, *13*, 1635, doi:10.3390/mi13101635 55

Peigen Zhou, Chen Wang, Jin Sun, Zhe Chen, Jixin Chen and Wei Hong
A 66–76 GHz Wide Dynamic Range GaAs Transceiver for Channel Emulator Application
Reprinted from: *Micromachines* **2022**, *13*, 809, doi:10.3390/mi13050809 67

Nathirulla Sheriff, Sharul Kamal, Hassan Tariq Chattha, Tan Kim Geok and Bilal A. Khawaja
Compact Wideband Four-Port MIMO Antenna for Sub-6 GHz and Internet of Things Applications
Reprinted from: *Micromachines* **2022**, *13*, 2202, doi:10.3390/mi13122202 81

Chuanba Zhang, Zhuoni Chen, Xiaojing Shi, Qichao Yang, Guiting Dong and Xuanhe Wei et al.
A Dual-Band Eight-Element MIMO Antenna Array for Future Ultrathin Mobile Terminals
Reprinted from: *Micromachines* **2022**, *13*, 1267, doi:10.3390/mi13081267 93

Shengjie Wang, Yuqi Zhao, Yishu Sun, Weicheng Wang, Jian Chen and Yang Zhang
Design of a Differential Low-Noise Amplifier Using the JFET IF3602 to Improve TEM Receiver
Reprinted from: *Micromachines* **2022**, *13*, 2211, doi:10.3390/mi13122211 107

Xiang Liu, Jiao Zhang, Shuang Gao, Weidong Tong, Yunwu Wang and Mingzheng Lei et al.
Demonstration of 144-Gbps Photonics-Assisted THz Wireless Transmission at 500 GHz Enabled by Joint DBN Equalizer
Reprinted from: *Micromachines* **2022**, *13*, 1617, doi:10.3390/mi13101617 123

Nazar Muhammad Idrees, Zijie Lu, Muhammad Saqlain, Hongqi Zhang, Shiwei Wang and Lu Zhang et al.
A W-Band Communication and Sensing Convergence System Enabled by Single OFDM Waveform
Reprinted from: *Micromachines* **2022**, *13*, 312, doi:10.3390/mi13020312 137

Yuqian He, Lu Zhang, Shanyun Liu, Hongqi Zhang and Xianbin Yu
Secure Transmission of Terahertz Signals with Multiple Eavesdroppers
Reprinted from: *Micromachines* **2022**, *13*, 1300, doi:10.3390/mi13081300 157

Editorial

Editorial for the Special Issue on Broadband Terahertz Devices and Communication Technologies

Lu Zhang [1,*], Xiaodan Pang [2,*] and Prakash Pitchappa [3,*]

1. College of Information Science and Electronic Engineering, Zhejiang University, Hangzhou 310027, China
2. Department of Applied Physics, KTH Royal Institute of Technology, 114 19 Stockholm, Sweden
3. Institute of Microelectronics, Agency for Science, Technology and Research, Singapore 138634, Singapore
* Correspondence: zhanglu1993@zju.edu.cn (L.Z.); xiaodan@kth.se (X.P.); prakash_pitchappa@ime.a-star.edu.sg (P.P.)

Citation: Zhang, L.; Pang, X.; Pitchappa, P. Editorial for the Special Issue on Broadband Terahertz Devices and Communication Technologies. *Micromachines* 2023, 14, 1044. https://doi.org/10.3390/mi14051044

Received: 11 May 2023
Accepted: 11 May 2023
Published: 12 May 2023

Copyright: © 2023 by the authors. Licensee MDPI, Basel, Switzerland. This article is an open access article distributed under the terms and conditions of the Creative Commons Attribution (CC BY) license (https://creativecommons.org/licenses/by/4.0/).

The remarkable explosion of wireless devices and bandwidth-consuming Internet applications have boosted the demand for wireless communications with ultra-high data rate. The wireless traffic volume is foreseen to match or even surpass the wired services by 2030, and high-precision wireless services will need to be guaranteed with a peak data rate of well beyond 100 Gbit/s, eventually reaching 1 Tbit/s. To meet the exponentially increasing traffic demand, new regions in the radio spectrum are being explored. The terahertz band, which is sandwiched between microwave frequencies and optical frequencies, is considered a next breakthrough point to revolutionize communication technology due to its rich spectrum resources. It is recognized as a promising candidate for future rate-greedy applications, such as 6G communications. At the World Radio Communication Conference 2019 (WRC-19), it was announced that the identification of frequency bands in the frequency range of 275 GHz–450 GHz is permitted for land-mobile and fixed service applications, indicating potential standardization of the low-frequency window of terahertz band for near-future wireless communications.

Motivated by the potential of terahertz wireless communications, this Special Issue reports on recent critical technological breakthroughs in terms of broadband terahertz devices and communications, as well as novel technologies at other frequency bands that can also motivate terahertz research. Five studies [1–5] present key devices for terahertz communications, including terahertz reconfigurable intelligent surfaces [1], terahertz microelectro-mechanical system (MEMS) switches [2], resonant triple-band terahertz thermal detectors [3], G-band continuous-wave traveling wave tubes [4], and wide-dynamic-range GaAs transceivers [5], which could effectively support broadband terahertz systems. Furthermore, we have also selected three interesting research studies [6–8] on low-frequency bands for this Special Issue, including the design of 5G multiple-input multiple-output (MIMO) antennas [6,7] and differential low-noise amplifiers [8]. We believe these works could also motivate research on terahertz communication devices and systems for 6G communications and other typical application scenarios. With the advances in broadband terahertz devices and the design of novel digital signal-processing routines, high-speed terahertz communications could be realized. In this Special Issue, three terahertz communicating systems were analyzed and demonstrated [9–11], including a 144 Gbps photonics terahertz communication system working at 500 GHz [9], a W-band communication and sensing convergence system [10], and an analysis of secure terahertz communications with perfect electric conductor (PEC) and multiple eavesdroppers [11].

To overcome the high loss and line-of-sight connectivity challenges of terahertz communication links, reconfigurable intelligent surfaces (RISs) are widely analyzed. However, active elements used for 5G RIS are often impractical for future 6G communications due to cutoff frequency limitations and higher loss at terahertz frequencies. Yang et al. [1] provided a comprehensive review on reconfigurable metasurfaces operating in the terahertz band with the potential to assist 6G communication links, categorized based on tuning mechanisms, including complementary metal-oxide-semiconductor (CMOS) transistors, Schottky diodes, high-electron mobility transistors (HEMTs), graphene, photoactive semiconductor materials, phase-change materials (vanadium dioxide, chalcogenides, and liquid crystals), and MEMS. Terahertz RISs are believed to be crucial for actualizing 6G communication links.

MEMS switches are important elements for future terahertz communication networks, but designing them for the terahertz frequency band is challenging since their physical dimensions are comparable to the wavelength. Feng et al. [2] designed and realized a terahertz MEMS switch using both silicon and fused quartz as examples of high and low dielectric-constant substrates, respectively. Both silicon and fused-quartz switches were calibrated based on the two-port through-reflection-line method from 140 to 750 GHz. At 750 GHz, the measurement results from the switches on both substrates show an ON-state insertion loss of less than 3 dB and an OFF-state isolation greater than 12 dB.

Terahertz waves possess unique properties, such as the ability to penetrate non-conductive materials and identify specific materials based on their characteristic terahertz signatures. Thus, terahertz detectors show great application potential in imaging, spectroscopy, and sensing fields. Wang et al. [3] experimentally validated a room-temperature CMOS monolithic resonant triple-band terahertz thermal detector. The responsivity, noise equivalent power, and thermal time constant of the detector were experimentally assessed at 0.91 THz, 2.58 THz, and 4.2 THz. The detector also has natural scalability to focal plane arrays, demonstrating significant advances in developing compact, room-temperature, low-cost, and mass-production multiband terahertz detection systems.

At the terahertz band, the G-band electromagnetic wave provides availability for the design of terrestrial and satellite radio communication networks according to the radio regulations of the International Telecommunication Union. The European Commission Horizon 2020 ULTRAWAVE project aims to exploit portions of the millimeter-wave spectrum for creating a very high-capacity layer. Feng et al. [4] presented the development of a G-band broadband continuous-wave traveling wave tube for wireless communications based on a slow-wave structure of fold waveguide. The device successfully provides a saturation output power over 8 W and a saturation gain over 30.5 dB with a bandwidth of 27 GHz.

Due to the abundance of spectrum resources in the millimeter-wave band, the WRC-19 conference approved multiple mm-wave spectra for future mobile communication research and development, including the 66–67 GHz frequency range, which is near the terahertz band. Zhou et al. [5] presented a 66–67 GHz transceiver monolithic microwave-integrated circuit (MMIC) in a waveguide module for massive MIMO channel emulator applications. The proposed transceiver integrates a tripler chain for local oscillator drive, a mixer, and a band-pass filter using a 0.1 µm pHEMT GaAs process. A high dynamic output power range, up to 50 dB over 66-to-76 GHz, is achieved by dealing with all unwanted harmonic signals employing highly selective band-pass and high-pass filters in the transceiver. The total power consumption of the chip is 645 mW with a supply voltage of 5 V.

In addition to the aforementioned research works, we have also selected three interesting studies [6–8] at low-frequency bands for this Special Issue, including the design of 5G MIMO antennas by Sheriff et al. [6] and Zhang et al. [7], and differential low-noise amplifiers by Wang et al. [8]. We believe these studies could also motivate research on terahertz communication devices and systems for 6G communications and other typical application scenarios.

Supported by broadband terahertz devices, terahertz communication systems have developed rapidly in recent years. Terahertz wireless communication systems based on photonics have emerged as promising candidates for 6G communications, capable of providing hundreds of Gbps or even Tbps data capacity. Liu et al. [9] experimentally demonstrated a 144 Gbps dual-polarization quadrature-phase-shift-keying (DP-QPSK) signal generation and transmission over a 20 km SSMF and 3 m wireless 2×2 MIMO link. A novel and low-complexity joint deep belief network (J-DBN) equalizer was proposed to compensate for linear and nonlinear distortions during fiber–wireless transmission. This scheme shows promises for meeting future fiber–terahertz integration communication demands for low power consumption, low cost, and high capacity.

The convergence of communication and sensing is highly desirable for future wireless systems. Idrees et al. [10] presented a converged system using a single orthogonal frequency-division multiplexing (OFDM) waveform and proposed a novel method, based on the zero-delay shift for received echoes, to extend the sensing range beyond the cyclic prefix interval (CPI). Both simulation and proof-of-concept experiments evaluated the performance of the proposed system at the W-band (97 GHz). The experiment employed a W-band heterodyne structure to transmit/receive an OFDM waveform featuring 3.9 GHz bandwidth with quadrature amplitude modulation (16-QAM). The proposed approach successfully achieves a range resolution of 0.042 m and a speed resolution of 0.79 m/s with an extended range, revealing the potential of terahertz technologies in the field of communication and sensing convergence.

The transmission security of high-speed THz wireless links is an important issue in terahertz research. He et al. [11] comprehensively investigated the physical layer security issue of a terahertz communication system in the presence of multiple eavesdroppers and beam scattering. The method of moments (MoM) was adopted to characterize the eavesdroppers' channel. To establish a secure link, traditional beamforming and artificial noise (AN) beamforming were considered as transmission schemes for comparison. The numerical results show that eavesdroppers can indeed degrade the secrecy performance by changing the size or location of the PEC, while the AN beamforming technique can be an effective candidate to counterbalance this adverse effect.

We would like to thank all the authors for their submissions to this Special Issue. We also thank the reviewers for dedicating their time and helping to ensure the quality of the submitted papers. Finally, we are grateful to the staff at the Editorial Office of *Micromachines*, and in particular Mr. Scott Wang, for their efficient assistance.

Conflicts of Interest: The authors declare no conflict of interest.

References

1. Yang, F.; Pitchappa, P.; Wang, N. Terahertz Reconfigurable Intelligent Surfaces (RISs) for 6G Communication Links. *Micromachines* **2022**, *13*, 285. [CrossRef] [PubMed]
2. Feng, Y.; Tsao, H.; Barker, N. THz MEMS Switch Design. *Micromachines* **2022**, *13*, 745. [CrossRef] [PubMed]
3. Wang, X.; Li, T.; Yan, S.; Wang, J. Room-Temperature CMOS Monolithic Resonant Triple-Band Terahertz Thermal Detector. *Micromachines* **2023**, *14*, 627. [CrossRef] [PubMed]
4. Feng, Y.; Bian, X.; Song, B.; Li, Y.; Pan, P.; Feng, J. A G-Band Broadband Continuous Wave Traveling Wave Tube for Wireless Communications. *Micromachines* **2022**, *13*, 1635. [CrossRef] [PubMed]
5. Zhou, P.; Wang, C.; Sun, J.; Chen, Z.; Chen, J.; Hong, W. A 66–76 GHz Wide Dynamic Range GaAs Transceiver for Channel Emulator Application. *Micromachines* **2022**, *13*, 809. [CrossRef] [PubMed]
6. Sheriff, N.; Kamal, S.; Tariq Chattha, H.; Kim Geok, T.; Khawaja, B. Compact Wideband Four-Port MIMO Antenna for Sub-6 GHz and Internet of Things Applications. *Micromachines* **2022**, *13*, 2202. [CrossRef] [PubMed]
7. Zhang, C.; Chen, Z.; Shi, X.; Yang, Q.; Dong, G.; Wei, X.; Liu, G. A Dual-Band Eight-Element MIMO Antenna Array for Future Ultrathin Mobile Terminals. *Micromachines* **2022**, *13*, 1267. [CrossRef] [PubMed]
8. Wang, S.; Zhao, Y.; Sun, Y.; Wang, W.; Chen, J.; Zhang, Y. Design of a Differential Low-Noise Amplifier Using the JFET IF3602 to Improve TEM Receiver. *Micromachines* **2022**, *13*, 2211. [CrossRef] [PubMed]
9. Liu, X.; Zhang, J.; Gao, S.; Tong, W.; Wang, Y.; Lei, M.; Hua, B.; Cai, Y.; Zou, Y.; Zhu, M. Demonstration of 144-Gbps Photonics-Assisted THz Wireless Transmission at 500 GHz Enabled by Joint DBN Equalizer. *Micromachines* **2022**, *13*, 1617. [CrossRef] [PubMed]
10. Idrees, N.; Lu, Z.; Saqlain, M.; Zhang, H.; Wang, S.; Zhang, L.; Yu, X. A W-Band Communication and Sensing Convergence System Enabled by Single OFDM Waveform. *Micromachines* **2022**, *13*, 312. [CrossRef] [PubMed]
11. He, Y.; Zhang, L.; Liu, S.; Zhang, H.; Yu, X. Secure Transmission of Terahertz Signals with Multiple Eavesdroppers. *Micromachines* **2022**, *13*, 1300. [CrossRef] [PubMed]

Disclaimer/Publisher's Note: The statements, opinions and data contained in all publications are solely those of the individual author(s) and contributor(s) and not of MDPI and/or the editor(s). MDPI and/or the editor(s) disclaim responsibility for any injury to people or property resulting from any ideas, methods, instructions or products referred to in the content.

Review

Terahertz Reconfigurable Intelligent Surfaces (RISs) for 6G Communication Links

Fengyuan Yang, Prakash Pitchappa * and Nan Wang *

Institute of Microelectronics, Agency for Science, Technology and Research, Singapore 138634, Singapore; yang_fengyuan@ime.a-star.edu.sg
* Correspondence: prakash_pitchappa@ime.a-star.edu.sg (P.P.); wangn@ime.a-star.edu.sg (N.W.)

Abstract: The forthcoming sixth generation (6G) communication network is envisioned to provide ultra-fast data transmission and ubiquitous wireless connectivity. The terahertz (THz) spectrum, with higher frequency and wider bandwidth, offers great potential for 6G wireless technologies. However, the THz links suffers from high loss and line-of-sight connectivity. To overcome these challenges, a cost-effective method to dynamically optimize the transmission path using reconfigurable intelligent surfaces (RISs) is widely proposed. RIS is constructed by embedding active elements into passive metasurfaces, which is an artificially designed periodic structure. However, the active elements (e.g., PIN diodes) used for 5G RIS are impractical for 6G RIS due to the cutoff frequency limitation and higher loss at THz frequencies. As such, various tuning elements have been explored to fill this THz gap between radio waves and infrared light. The focus of this review is on THz RISs with the potential to assist 6G communication functionalities including pixel-level amplitude modulation and dynamic beam manipulation. By reviewing a wide range of tuning mechanisms, including electronic approaches (complementary metal-oxide-semiconductor (CMOS) transistors, Schottky diodes, high electron mobility transistors (HEMTs), and graphene), optical approaches (photoactive semiconductor materials), phase-change materials (vanadium dioxide, chalcogenides, and liquid crystals), as well as microelectromechanical systems (MEMS), this review summarizes recent developments in THz RISs in support of 6G communication links and discusses future research directions in this field.

Keywords: reconfigurable intelligent surface (RIS); terahertz (THz); 6G communication; reconfigurable metasurface

1. Introduction

With the commercial launch of the fifth generation (5G) network in 2020, research into the sixth generation (6G) communication system is eagerly on the agenda [1–3]. In order to meet the communication requirements of modern society in building smart cities, the 6G wireless network is envisioned to provide ultra-fast data rates (~1 Tbps), ultra-low latency (<1 ms), ubiquitous wireless connectivity, superior spectral and energy efficiency, as well as extremely high reliability and security [4–6]. The terahertz (THz) spectrum, ranging from 0.1 THz to 10 THz, is well suited for 6G applications due to its higher frequency, large bandwidth, and short response time [7–9]. However, the THz spectrum is highly absorbed by water molecules in the atmosphere, and its ultra-short wavelength is easily scattered by obstacles along the propagation path. This high path loss limits the THz wave to short-range transmission. Energy-concentrated directive beam transmission is imperative for the THz network to compensate for propagation losses. In contrast with active phased-array antennas [10,11] applied on the transmitting and receiving end, reconfigurable intelligent surfaces (RISs) are proposed to modify the wireless transmission path [12–16]. RIS consists of periodic elements (meta-atoms or unit cells) altering the phase and magnitude of incident waves to have constructive/destructive interference in desired directions. Based on this, THz wireless communication is capable of being extended to non-line-of-sight (NLOS) transmission (Figure 1a) [17]. Compared with existing relays using power amplifiers for wave receiving and retransmitting, RIS enhances the transmission by phase control using nearly passive elements, requiring less spectrum and energy [4,13]. Therefore, RISs can be embedded within infrastructure cost-effectively, enabling advanced wireless communication in 6G smart cities (Figure 1b) [18]. RIS is also known as reprogrammable intelligent surface [19,20] or intelligent reflecting surface (IRS) [21,22] in some contexts. The latter primarily denotes an intelligent surface operating in the reflective mode. A promising application of such a surface is deploying it on infrastructure to reflect waves in a desired direction [4].

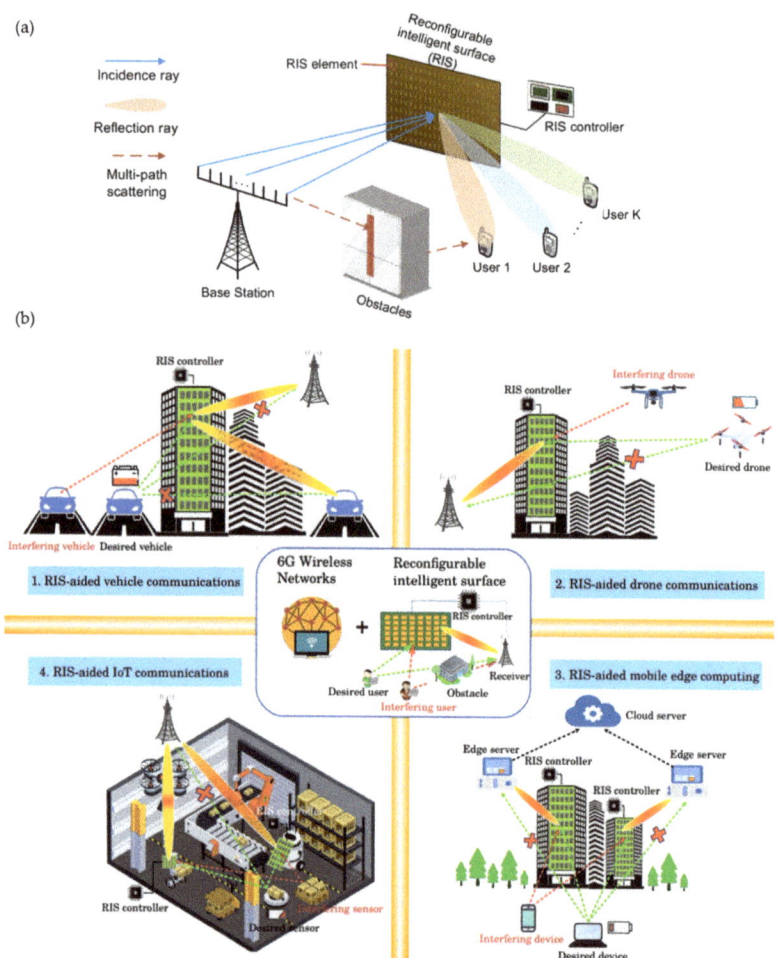

Figure 1. RIS-assisted wireless communication. (**a**) RIS-enabled non-line-of-sight (NLOS) transmission. Reprinted from Ref. [17]. (**b**) RIS-embedded smart infrastructures for future 6G communications. Reprinted from Ref. [18].

RIS benefits have been realized as a result of the flourishing development of electromagnetic metasurfaces in the last decade. Metasurfaces are two-dimensional versions of metamaterials and have the advantages of low profile, reduced loss, and ease of fabrication [3,4]. Metasurfaces consist of artificially designed periodic structures, enabling unprecedented wave manipulation that is unattainable with natural materials [23]. In

2011, Yu et al. first proposed the generalized Snell's law, including abrupt phase shifts from the metasurface unit cells to realize nearly arbitrary wavefront shaping [24]. Since then, various fascinating wavefront manipulations have been demonstrated, such as reflectionless wide-angle refraction [25,26], sub-diffraction focusing with ultrathin planar lens [27,28], and wave-impedance matching across different material boundaries [29]. Such metasurfaces have only a static function once fabricated, whereas reconfigurable metasurfaces with tunable functions are gaining attention. The reconfigurable function is achieved by incorporating meta-atoms with tunable components or materials, which alter the electromagnetic response through external stimuli. With the development of digital control, coding metasurfaces were proposed to manipulate waves by changing the coding sequences of digital particles (unit cells) [30–36]. Active electronic components, such as positive-intrinsic-negative (PIN) diodes [30,37–39], have been widely applied in RISs working at 5G networks but are incompetent for applications in 6G systems due to the cutoff frequency limit. To overcome this challenge, reconfigurable technology approaches from both the electronic and optical sides are being explored intensively to fill the "THz gap" [40]. Beyond that, phase-change materials and microelectromechanical systems (MEMS) for THz RISs are also widely investigated.

Most existing review papers focused on summarizing various possible applications of RISs in wireless communication [41] or analyzing different tunable metasurfaces according to their tuning mechanisms [42,43]. Several review papers analyzed active metasurfaces for only one tuning element [44–47]. This paper emphasizes reconfigurable metasurfaces operating in the THz spectrum with the potential to assist 6G communications, i.e., metasurfaces with tunable functions such as pixel-level amplitude modulation or wide-range phase coverage for beam steering. Polarization modulation is briefly discussed as a possible future carrier-wave function. To underline the reconfigurability, RISs are categorized based on their tuning mechanisms, including complementary metal-oxide-semiconductor (CMOS) transistors, Schottky diodes, high-electron mobility transistors (HEMTs), graphene, photoactive semiconductor materials, phase-change materials (vanadium dioxide, chalcogenides, and liquid crystals), and MEMS.

The rest of this paper is organized as follows: Section 2 presents various reconfigurable metasurfaces with pixel-level amplitude modulation or dynamic-beam-steering functions at THz frequency (0.1 THz–10 THz). Subsections are divided according to the tuning elements. Section 3 summarizes the properties of THz RISs and discusses possible future research directions.

2. Terahertz Reconfigurable Intelligent Surface

Various THz RISs with the potential to assist 6G communications are discussed in this section. The functionality of the reconfigurable metasurface focuses on amplitude modulation at the pixel-level or phase modulation for dynamic beam steering. The subsections are devoted to the tuning elements, investigating the reconfigurable mechanism for achieving RIS. The analysis begins with electronic approaches (CMOS transistors, Schottky diodes, HEMTs, and graphene), followed by optical approaches involving semiconductor materials, then phase change materials (vanadium dioxide, chalcogenides, and liquid crystals), and finally MEMS-based structural deformation.

2.1. Electronic Approaches

Active electronic components, such as diodes and varactors, have been widely utilized in microwave and millimeter-wave regimes for reconfigurable metasurface control [30,37–39,48–51]. However, they are incompatible with 6G applications in the THz spectrum due to the restricted cutoff frequency and dramatically increased loss [52]. To overcome this obstacle and implement reconfigurable metasurfaces in the THz spectrum using electronic approaches, CMOS transistors have to be embedded in meta-atoms with specially designed structures for local resonances [52]. Alternative solutions include integrating metasurface meta-atoms with layered semiconductor structures or graphene with electrical tunability.

2.1.1. Complementary Metal-Oxide-Semiconductor (CMOS) Transistor

Incorporating active electronic devices into a passive metasurface allows for fast dynamic control, but the application is limited to microwave frequencies. In 2020, Venkatesh et al. presented a programmable metasurface using complementary metal-oxide-semiconductor (CMOS) transistors operating at 0.3 THz beyond its cutoff frequencies [52]. They applied a 65 nm industry-standard CMOS process to fabricate the metasurface in a silicon chip tile, consisting of 12 × 12 arrays (Figure 2a). Each meta-atom contained eight n-type metal-oxide-semiconductor (NMOS) transistors for an eight-bit reconfiguration at gigahertz (GHz) speed. Parallel subwavelength inductive microloops were added to the transistor for local resonance to suppress the non-negligible parasitic capacitance leakage. An amplitude modulation of 25 dB was achieved between all switches open (maximum transmission) and closed (minimum transmission) states. A demonstration of the 2 × 2 tiled chips (Figure 2b) for holographic projections of the letter 'P' is shown in Figure 2c. The unit-cell structure was derived from a general C-shaped split-ring resonator (SRR), which controls the transmitted amplitude and phase by varying the gap-opening orientation and size, respectively. This was realized by selectively switching the eight transistors of a meta-atom, leading to 256 states in total. Due to the symmetric configuration, each meta-atom had 84 unique codes and realized a phase coverage of 260°. Beam steering from 0° to ±30° was achieved by configuring the unit cell in three different phase profiles and meta-atom digital settings (Figure 2d).

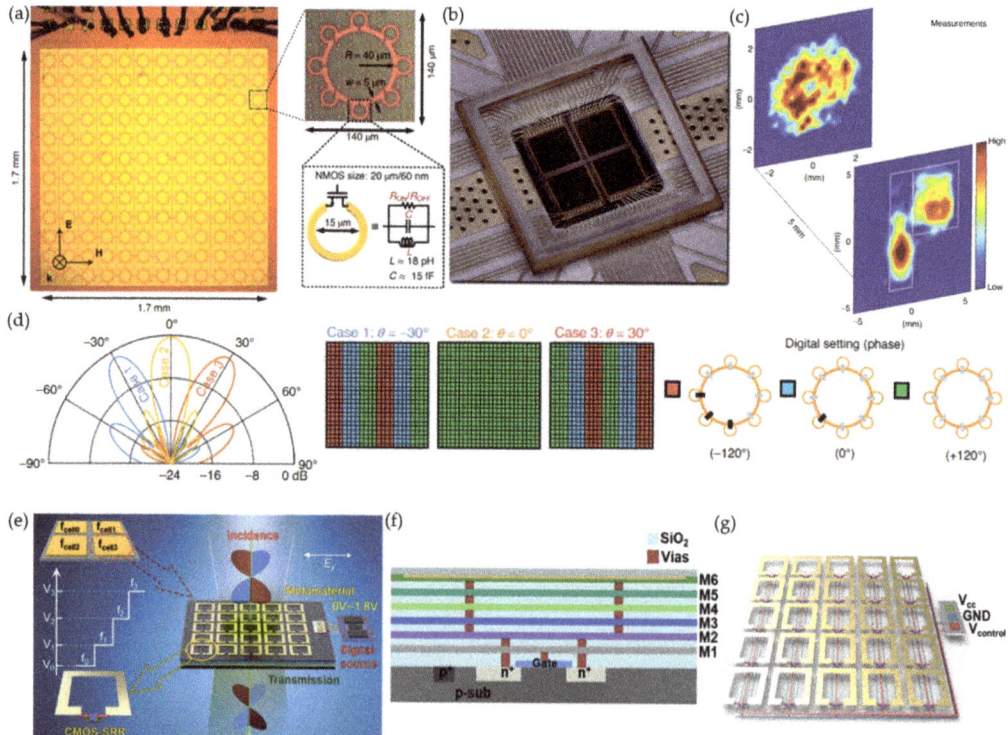

Figure 2. CMOS transistor-enabled reconfigurable metasurface. (**a–d**) GHz-speed programmable metasurfaces using CMOS-based chip tiles. (**a**) A single silicon chip tile consists of a 12 × 12 array (left).

The enlarged portion (right) shows the unit-cell structure with active NMOS transistors embedded in the gap of inductive microloops. Each unit cell has an eight-bit control, enabling 256 states for amplitude and phase control. (b) Photo of the fabricated 2 × 2 tiled chips, which were wire-bonded to a customized printed circuit board for external voltage control. (c) Amplitude modulation was experimentally demonstrated as a holographic projection of the letter 'P'. (d) Beam steering at ±30° with the corresponding three different phase profiles and meta-element digital settings. Reprinted from Ref. [52]. (e–g) Reconfigurable metasurface based on CMOS structures. (e) A bias voltage is applied for transmitted amplitudes and phase modulation. The reconfigurable metamaterial can be divided into subsections for greater functionality. (f) Cross section of the unit cell, consisting of six layers for the CMOS transistor configuration. (g) Layout with wire connection for the biasing control. Reprinted from Ref. [53].

By utilizing custom-designed CMOS-based semiconductor structures, it is possible to circumvent the cutoff frequency limitation of commercial transistors (Figure 2e) [53]. The meta-atom structure, consisting of a square SRR on top of a six-layer CMOS-based semiconductor structure, is shown in Figure 2f. The gap of the SRR was connected to the source and drain of a metal-oxide-semiconductor field-effect transistor (MOSFET) through vias (Figure 2g). The metasurface was fabricated with 180 nm CMOS technology. With a bias voltage from 0 V to 1.8 V, a redshifted frequency of 35 GHz and a phase difference of 3° were achieved at 0.3 THz. Although the phase modulation was limited, the authors presented an alternative solution for realizing a CMOS-based reconfigurable metasurface.

2.1.2. Schottky Diode

Commercial off-the-shelf (COTS) diodes have been extensively used in 5G RIS for active tuning; the concept was adopted in the THz spectrum by constructing semiconductor metamaterials with a Schottky gate structure. This idea was first realized by an active metamaterial switch in 2006 [54]. The Schottky junction was formed by integrating a metallic SRR with a 1 µm thick n-doped gallium arsenide (GaAs) layer (Figure 3a) [55]. Applying a reverse gate-bias voltage alters the substrate charge-carrier density around the split gap, thus affecting the resonance response of the SRRs. Later, the same group used the unit cell to build a 4 × 4 pixelated spatial light modulator (Figure 3b). Each pixel, consisting of 50 × 50 split-ring resonators (SRRs) with a total size of 4 × 4 mm^2, was independently controlled by the external bias voltage. An amplitude-modulation depth of ~3 dB was achieved at 16 V bias voltage in kilohertz (kHz) speed. The modulation speed was enhanced to megahertz (MHz) by placing the ohmic ground plane directly underneath the Schottky layer to minimize the device capacitance, as shown in Figure 3c [56]. This was utilized to build a four-color, 8 × 8 pixelated spatial light modulator by repeating a 2 × 2 four-color subarray, thereby realizing a more advanced spatial and spectral control (Figure 3d).

To realize beam steering, a switchable-diffraction grating with combined amplitude and phase modulation using Schottky gate structure for tuning was presented in [57] (Figure 3e). By applying a reverse bias voltage (−13 V) on alternate columns, the metasurface achieved 20 dB amplitude modulation at 36.1° with a speed of 1 kHz at 0.4 THz. An alternative method applied to achieve beam steering was shown using meta-atoms with specially designed 'C'-shaped structures covering a 2π phase range. Figure 3f shows eight 'C'-shaped SRRs with different gap sizes and orientations for a $\pi/4$ phase gradient [58]. The metasurface metallic pattern was made using gold film and embedded with a doped semiconductor substrate (Figure 3g) [58]. As a result, the metasurface realized broadband (0.55 to 0.83 THz) beam steering with a deflection angle from 59.09° to 34.88° at a modulation speed of 3 kHz (Figure 3h).

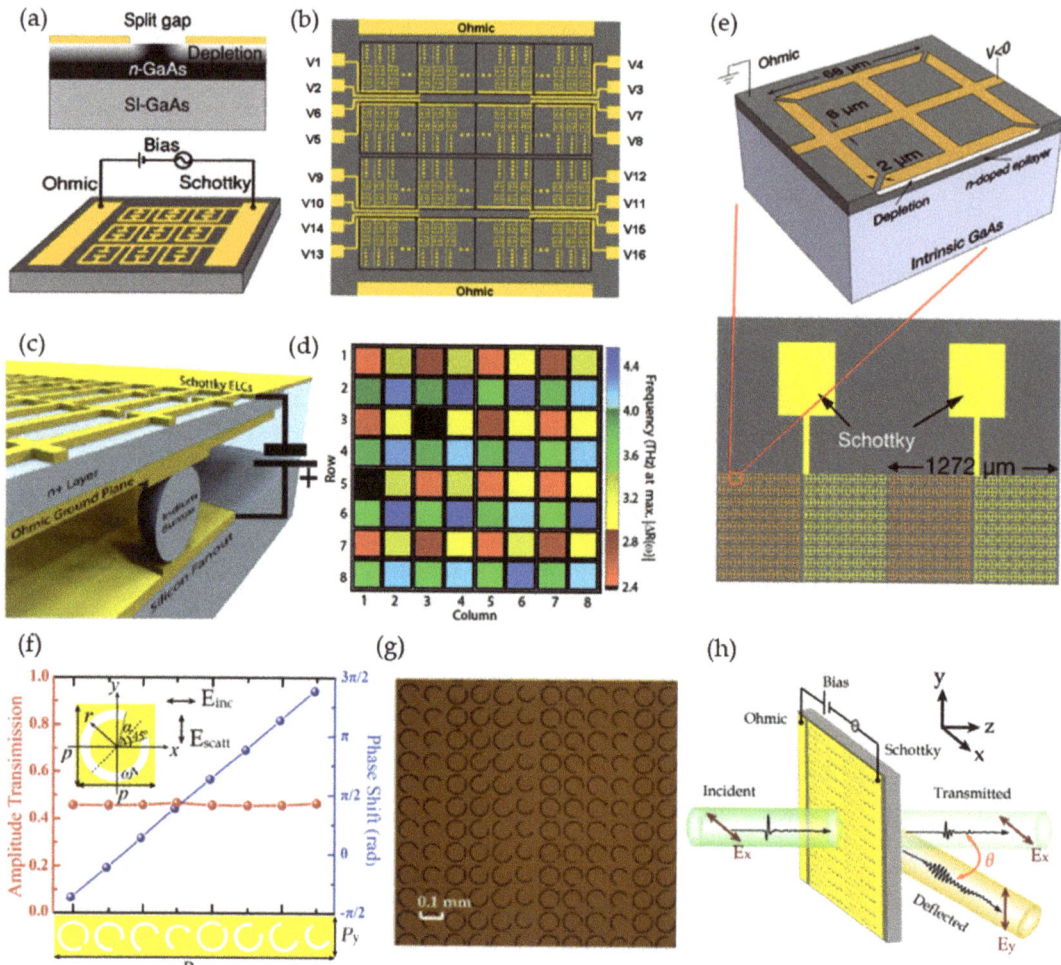

Figure 3. Schottky-diode-structure-enabled reconfiguration. (**a**,**b**) A 4 × 4 pixel amplitude modulator. (**a**) Schematic of the cross section of the unit cell incorporating SRR with Schottky gate structure (top). The gray scale of the depletion region indicates the free charge-carrier density. A single pixel on the THz SLM for amplitude modulation (bottom). (**b**) THz SLM consisting of 4 × 4 pixels. Reprinted from Ref. [55]. (**c**,**d**) An 8 × 8 four-color spatial light modulator. (**c**) Schematic of the metamaterial absorber with a flip-chip-bonded, n-doped GaAs epitaxial layer. (**d**) An example of the spatial light modulator with different frequencies for each pixel. Reprinted from Ref. [56]. (**e**) A diffractive modulator with grating configuration realizing 22 dB amplitude modulation at 36.1°. Reprinted from Ref. [57]. (**f**–**h**) A phase-modulated deflector. (**f**) An array consisting of eight unit cells realized 2π phase control with nearly the same transmission efficiency. (**g**) Microscopic image of the fabricated metasurface. (**h**) An illustration of the deflected wave transmission. Reprinted from Ref. [58].

2.1.3. High-Electron Mobility Transistor (HEMT)

The instability of two-dimensional electron gases (2DEGs) in short-channel high-electron mobility transistors (HEMTs) leads to a resonant response at the geometrical plasmon frequency, which depends on the size and shape of the channel [59–61]. A pseudomorphic HEMT-integrated metadevice was first introduced in 2011 by D. Shrekenhamer et al.

(Figure 4a) [61]. HEMTs were integrated beneath the capacitive gaps of a square electric-LC (ELC) resonator. The resonance response was reconfigured by changing the channel-carrier density through external bias voltage (1 V). The metadevice was fabricated using a commercial GaAs process and realized a modulation depth of 33% at 0.46 THz with a rate of 10 MHz. In 2016, the authors adopted the same meta-atom to demonstrate a 2 × 2 spatial light modulator (Figure 4b) [62].

Various structures have been designed to explore 2DEGs in HEMTs [63–65]. One such embedded HEMT with metal–insulator–metal (MIM) capacitors for amplitude modulation (45%) was introduced in [63] through a biasing voltage of 3 V at 0.58 THz. Reconfigurability with both amplitude and phase modulation is desirable. Double-channel heterostructures with two split channels of decreased polarized-carrier concentration were designed to support a nanoscale 2DEG layer with high concentration and mobility [66] (Figure 4c). An equivalent collective dipolar array was combined with a double-channel heterostructure. An external electrical signal was applied to control the electron concentration in the 10 nm thick 2DEG layers, which led to a resonant mode conversion between two dipolar resonances, providing fast amplitude and phase modulations. Depletion of the 2DEG layer shifted the dipolar resonance from the long central wire to the short one, resulting in a blueshift of the resonance frequency. This design demonstrated 1 GHz modulation speed for the first time and achieved 85% modulation depth (Figure 4d) and 68° phase shift (Figure 4e) at ~0.351 THz. By combining inductance–capacitance resonance and dipole resonance, an enhanced-resonance active HEMT metasurface was designed (Figure 4f), realizing a phase modulation of 137° at 0.35 THz with a biasing voltage of 8 V (Figure 4g) [67].

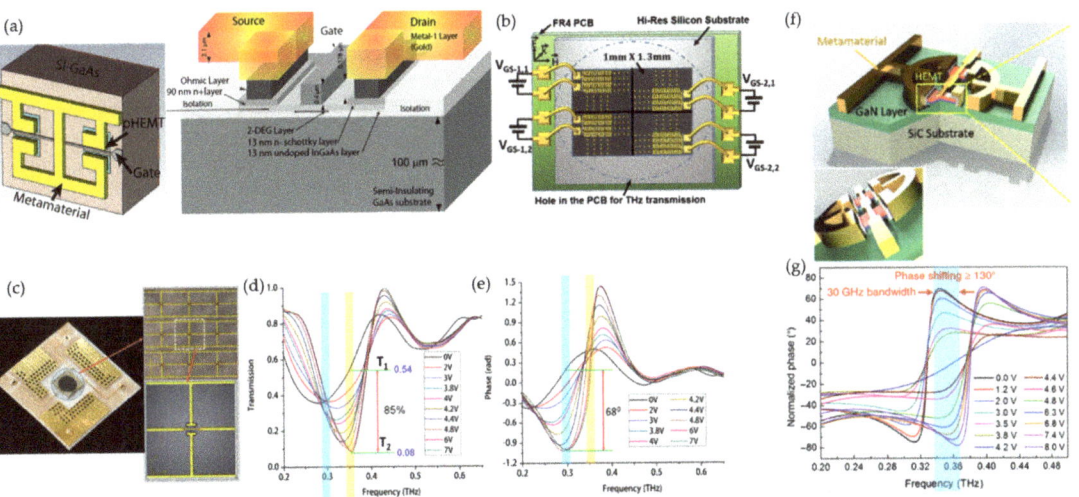

Figure 4. High-electron-mobility-transistor (HEMT)-enabled reconfigurable metasurface. (**a**,**b**) An HEMT-embedded metamaterial modulator with a speed of 10 MHz. (**a**) A simulation unit-cell model with HEMT beneath each split gap. The cross-sectional view is shown on the right. Reprinted from Ref. [61]. (**b**) A 2 × 2 spatial light modulator with a modulation depth of 33% at 0.46 THz. Reprinted from Ref. [62]. (**c**–**e**) Metasurface with 1 GHz modulation speed combining a dipolar array with a double-channel heterostructure. (**c**) Image of the fabricated metasurface. (**d**) Depth modulation of 85%. (**e**) Phase modulation of 68°. Reprinted from Ref. [66]. (**f**,**g**) Large phase modulator with HEMT embedded and an enhanced-resonance metasurface. (**f**) Schematic of the unit-cell structure. (**g**) Phase modulation of more than 130°. Reprinted from Ref. [67].

2.1.4. Graphene

Graphene is a two-dimensional (2D) material of honeycomb structures formed by single-layer carbon atoms arranged in hexagonal lattices. Graphene has complex conductivity that supports the propagation of plasmonic modes at THz frequencies [68]. The surface conductivity can be efficiently controlled through a perpendicular-bias electric field that induces charge carriers to shift the graphene chemical potential (Fermi level) away from the Dirac point. Compared with conventional semiconductors, graphene has the attractive advantages of high electron mobility, considerable thermal conductivity, and strong mechanical ductility [69]. Hence, graphene has considerable potential to be applied in the THz frequency for dynamic wave control.

A graphene-based electroabsorption modulator was demonstrated by placing atomically thin graphene layer on top of a dielectric substrate with a reflective metal back gate (Figure 5a) [70,71]. The substrate thickness was designed to be an odd multiple of a quarter-wavelength of the incident wave to enhance the modulation depth. With a biasing voltage of -10 V, the Fermi level in graphene was tuned at the Dirac point, realizing the maximal reflectance. The carrier concentration increased with increased voltage, resulting in enhanced absorption. The graphene layer was patterned using O_2 plasma to demonstrate a 4×4 pixelated reflectance modulator (Figure 5b) [71]. An alternative approach using electrolyte gating, producing higher charge densities to build a spatial phase modulator, was presented in [72] (Figure 5c). Large electric fields were generated at the graphene–electrolyte interface, giving rise to charge accumulation over a large area without an electrical short circuit and removing the thickness control. In order to build the 16×16 pixelated modulator, graphene was grown by chemical-vapor deposition (CVD) on a large area. Selective control of a single pixel was achieved by voltage biasing through the corresponding column and row (Figure 5d).

A reflect array combined with graphene for dynamic beam steering was first proposed using a square graphene patch (Figure 5e) [68]. Due to the slow wave propagation in the plasmonic mode of graphene, a patch was designed with much smaller dimensions ($\lambda/10$, compared with conventional conductors of $\lambda/2$) for a resonance response. This reduced interelement spacing allows for more efficient wavefront manipulation. By tuning the chemical potential from 0 to 0.52 eV, a phase coverage of $300°$ was obtained at 1.3 THz with a fixed patch dimension. To have full $360°$ phase coverage for dynamic-gradient phase control, graphene was designed with resonant structures (Figure 5f) [73,74] or combined with resonant metallic patterns [75,76]. A graphene-based coding metasurface for beam splitting was proposed by controlling the Fermi level (Figure 5g) [77–81]. Such coding metasurfaces also have great potential to be applied to security systems for message transmission [81]. Spatially selective column-level tuning was presented with graphene embedded in SRR, resulting in four deflection angles ($5°$, $11°$, $17°$, and $23°$) at 1.05 THz (Figure 5h) [82]. Such graphene-based beam steering metasurfaces had only been tested in simulations. The first experimentally demonstrated design was illustrated in [83] (Figure 5i). The graphene strip was placed at the gap of a bowtie structure for field concentration. Each column was individually controlled, resulting in a maximum beam steering of $\pm 25°$ at 1 THz (Figure 5j).

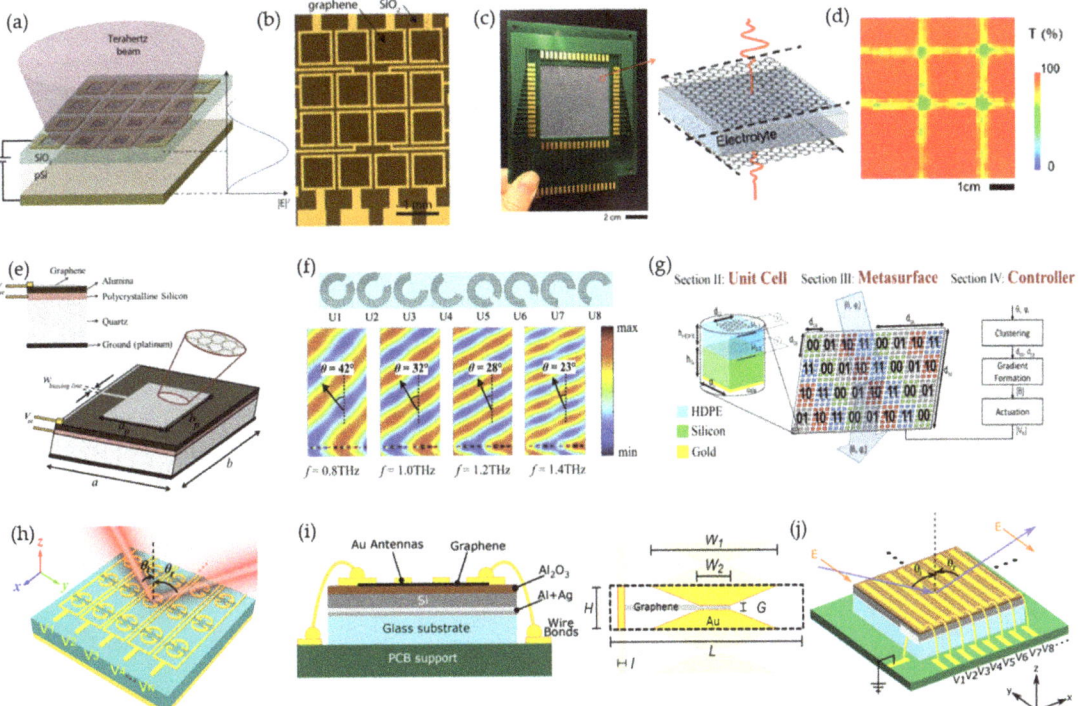

Figure 5. Graphene-based reconfigurable metasurface. (**a,b**) A 4 × 4 reflection modulator. (**a**) A schematic of the graphene-SiO$_2$-Si structures. The substrate has an optical thickness of an odd quarter wavelength. (**b**) Optical image of the graphene-enabled reflection modulator. Reprinted from Ref. [71]. (**c,d**) A 256-pixel spatial light modulator. (**c**) Photo of the modulator (left). The enlargement (right) shows the graphene–electrolyte–graphene unit-cell structure. (**d**) A THz transmission image at 0.1 THz with two rows and columns biased at +1.0 and −1.0 V, respectively. Reprinted from Ref. [72]. (**e**) Unit cell consisting of a square graphene patch for a tunable reflective metasurface at 1.3 THz. The cell has dimensions of a = b = 14 µm, $a_p = b_p$ = 10 µm. Reproduced with permission from [68]. (**f**) Beam steering with graphene patterned in SRRs. Reprinted from Ref. [73]. (**g**) Digital metasurface using a graphene–insulator–graphene stack for beam steering. Reprinted from Ref. [81]. (**h**) Column-level controlled beam steering with graphene embedded with SRR structures. Reprinted from Ref. [82]. (**i–j**) Experimentally demonstrated beam-steering metasurface with graphene embedded with a bowtie structure. (**i**) Cross section of the experimentally demonstrated beam-steering metasurface (left) and its unit-cell structure (right). (**j**) Schematic of the metasurface with the individually biased column. Reprinted from Ref. [83].

2.2. Optical Approaches

In photosensitive semiconductors (e.g., silicon, GaAs, and conducting oxide), conductivity can be controlled by pumping carriers from the valence band to the conduction band using an external laser beam with photon energy higher than that of the bandgap [84]. This dynamic photoconductivity provides temporal modulation of the metasurface. Figure 6a shows a hybrid circular SRR with an aluminum SRR placed on a circular silicon ring [84]. The top layer SRR was designed with eight patterns for 360° phase coverage, realizing a wideband (0.6–1 THz) cross-polarized wavefront deflection from 51° to 28° (Figure 6b). Stimulating the silicon with an external optical laser pump (800 nm) increased its conductivity and closed the SRR gap, eliminating the beam-splitting and deflection effect. A similar concept was applied to high-resistivity silicon resonators (Figure 6c) [85].

A supercell consisting of four resonators was designed, realizing a beam-deflection angle of 34.7° at 0.586 THz with 5 mW laser-pump power (Figure 6d). The rise time for the transient photocarrier was 14 ps. This symmetry-preserved Huygens' metasurface design achieved a high transmission efficiency of 90%.

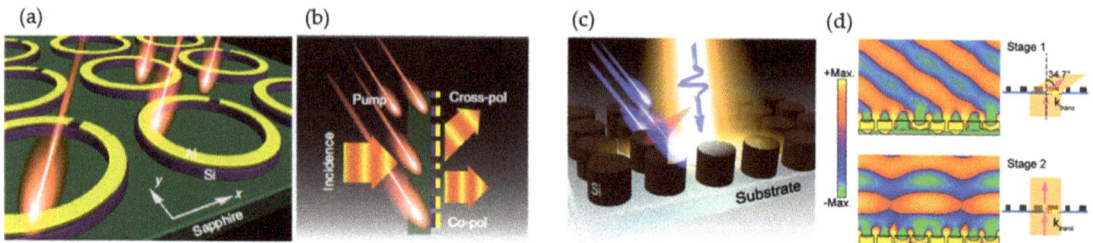

Figure 6. Semiconductor materials for temporal modulation. (**a,b**) Optical active polarization switching and dynamic beam splitting. (**a**) An illustration of the hybrid circular split-ring resonator (h-SRR) pumped by near-infrared femtosecond pulses. (**b**) An active polarizing beam splitter. Reprinted from Ref. [84]. (**c,d**) Spatiotemporal dielectric metasurfaces for beam steering. (**c**) Ultrafast femtosecond laser pulses (@ 800 nm, 100 fs) pump high-resistivity silicon on a quartz substrate, providing transient photocarriers for temporal modulation. (**d**) Temporal beam steering of 34.7° at 0.586 THz. Reprinted from Ref. [85].

2.3. Phase-Change Materials

Phase-change materials (PCMs) transform from the amorphous to the crystalline state or between different crystalline phases upon external stimulation [86]. Their reversible phase transition, thermal stability, and fast switching speed make PCMs extremely popular [87]. Various phase-change materials, such as vanadium dioxide (VO_2), chalcogenide phase-change materials, and liquid crystals (LCs), have been explored for reconfigurable metasurfaces.

2.3.1. Vanadium Dioxide (VO_2)

Vanadium dioxide (VO_2) has a unique atomic rearrangement of monoclinic phases, exhibiting semiconductor behavior at room temperature and rutile phases at high temperatures (~68°), experiencing an abrupt transition from insulator to metal when subjected to thermal, electrical, optical, or mechanical stimulation [88–92]. This insulator-to-metal transition property provided a diversified application for switching a metasurface between the broadband-absorber and half-wave plate states (Figure 7a) [90]. The metasurface had an absorption efficiency >90% at room temperature but became a reflector as the temperature rose above the phase-change temperature [90]. Supercells consisting of eight resonators were designed to have a beam-steering effect in the metallic state (Figure 7b). The metasurface had a simulated wideband deflection of ~28° at 0.8 THz (Figure 7c). Embedded VO_2 with resonant structures for a multifunctional coding metasurface was also presented numerically [91,93–95]. An experimental demonstration of VO_2-based reconfigurable beam steering was presented in [89] (Figure 7d). The unit cell had a cross-shaped aperture with a resistive heater electrode placed beneath for independent Joule-heating actuation. A 100 nm thick SiO_2 layer was used to insulate the resistive-heater electrodes from the top Au layer. The refractive index of the VO_2 was varied through temperature changes. The metasurface achieved a 44° beam deflection in both horizontal and vertical directions at 0.1 THz (Figure 7e,f). Recently, an electrically triggered VO_2-based reconfigurable metasurface for both phase (90°) and amplitude (71% at 0.79 THz) modulation was experimentally demonstrated in [96]. And a multi-state 8 × 8 pixelated reflective modulator was demonstrated with a modulation speed of 1 kHz [97].

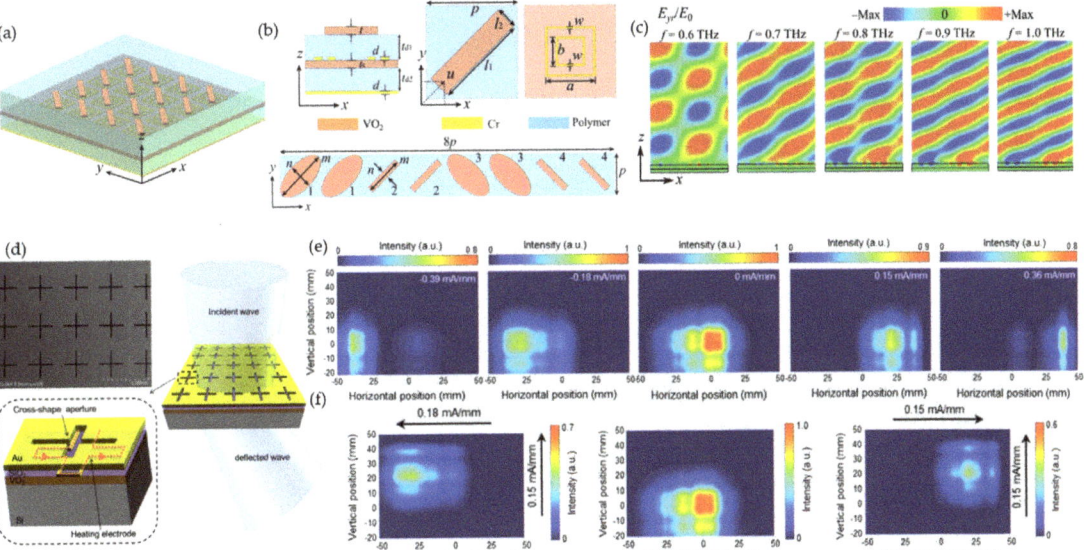

Figure 7. Phase-change materials of a vanadium-dioxide (VO$_2$)-enabled reconfigurable metasurface. (**a**–**c**) A thermally controlled reconfigurable metasurface with broadband absorption-to-reflection conversion. Reprinted from Ref. [90]. (**a**) Schematic of the VO$_2$ integrated metasurface. (**b**) Unit-cell structure (top) and a unit-cell array for 2π phase control (bottom). (**c**) Broadband reflection when VO$_2$ is in its fully metallic state. (**d**–**f**) An electronically controlled beam-steering metasurface operates at 0.1 THz. Reprinted from Ref. [89]. (**d**) Top-view scanning electron microscopic image of the metasurface (top left), unit-cell structure (bottom left), and beam steering (right). (**e**) Horizontal beam steering at −22°, −14°, 0°, 12°, and 22°. (**f**) Vertical beam steering at −14°, 0°, and 12°.

2.3.2. Chalcogenide Phase-Change Materials

Phase transitions of chalcogenide phase-change materials are mediated by nucleation dynamics, providing an analog response by continuously varying the crystallinity fraction. Moreover, these analog states are nonvolatile, requiring zero-hold power [86]. A chalcogenide phase-change material, germanium–antimony–tellurium (GST), a ternary compound made of germanium (Ge), antimony (Sb), and tellurium (Te), exhibits a reconfigurable phase-transition response between amorphous and crystalline states under external optical, electrical, or thermal stimulation. A multifunctional tunable metadevice was described in [98] using Ge$_2$Sb$_2$Te$_5$. A dual-split asymmetric SRR was designed for Fano resonance (Figure 8a). Four meta-atoms designed for different THz frequency responses were configured in a 2 × 2 array. A spatially selective reconfiguration was achieved by Joule heating from the isolated biasing current (Figure 8b). Continuously increasing the material temperature led to increased terahertz conductivity and refractive index, providing multilevel resonance modulation states (reaching 100% at 850 mA) (Figure 8c).

Germanium telluride (GeTe) with a crystallization temperature of ∼ 200 °C and a resistivity reduction of six orders has also been extensively studied for the construction of tunable metasurfaces [99,100]. Optical stimulus provides faster material phase-change modulation [98,101]. Recently, a two-bit coding metasurface based on GeTe was experimentally demonstrated with multifunctionality, including beam tilting, directing, and splitting at 0.3 THz (Figure 8d) [101]. A laser pulse excited the meta-atom between crystalline (conductive) and amorphous (insulating) states (Figure 8e), achieving a reflected phase difference of 180° (Figure 8f). Five different beam controls were demonstrated with five coding masks, as shown in Figure 8g.

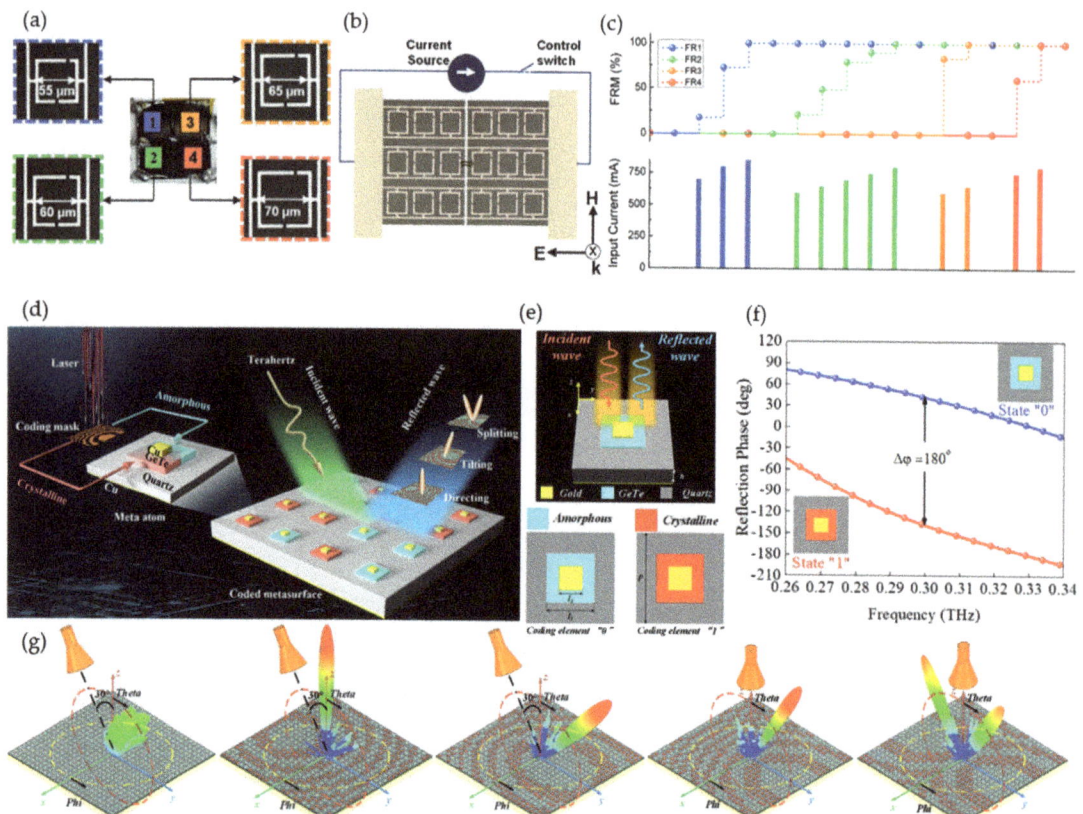

Figure 8. Chalcogenide phase-change materials enabled a reconfigurable metasurface. (**a**–**c**) Germanium–antimony–tellurium (GST) incorporated with Fano-resonance mate atoms for multicolor spatial light modulation. (**a**) A 2 × 2 array for four-color spatial light modulation. (**b**) Schematic of current biasing. (**c**) Multilevel Fano-resonance modulation (FRM) results from different input currents (stimulus period ≈ 15 s). Reprinted from Ref. [98]. (**d**–**g**) Phase-change GeTe material applied for a multifunctional coding metasurface. (**d**) Illustration of the coding metasurface. (**e**) GeTe- and gold-integrated unit cell with amorphous (insulating) state of GeTe and crystalline (conductive) state. (**f**) Reflected phase of 180° at 0.3 THz for the coding element at two different states. (**g**) Multifunctionality (beam tilting, directing, and splitting) is realized through different coding masks. Reprinted from Ref. [101].

2.3.3. Liquid Crystals

Liquid crystals are attractive for their inherent birefringent properties, which depend on the orientation of liquid crystal molecules and can be effectively controlled by an external electric field or light [102–105]. Figure 9a shows a THz spatial light modulator based on liquid crystals combined with metamaterial absorbers [102]. The authors used an isothiocyanate-based liquid-crystal mixture to fill the space around the electric ring resonator (ERR). The spatial light modulator consisted of a 6 × 6 pixel array (Figure 9b), and each pixel was individually controlled by a 15 V peak-to-peak square waveform at 1 kHz. The applied electric field forced the liquid-crystal molecule to align with its direction, achieving 75% reflectivity modulation at 3.67 THz (Figure 9c).

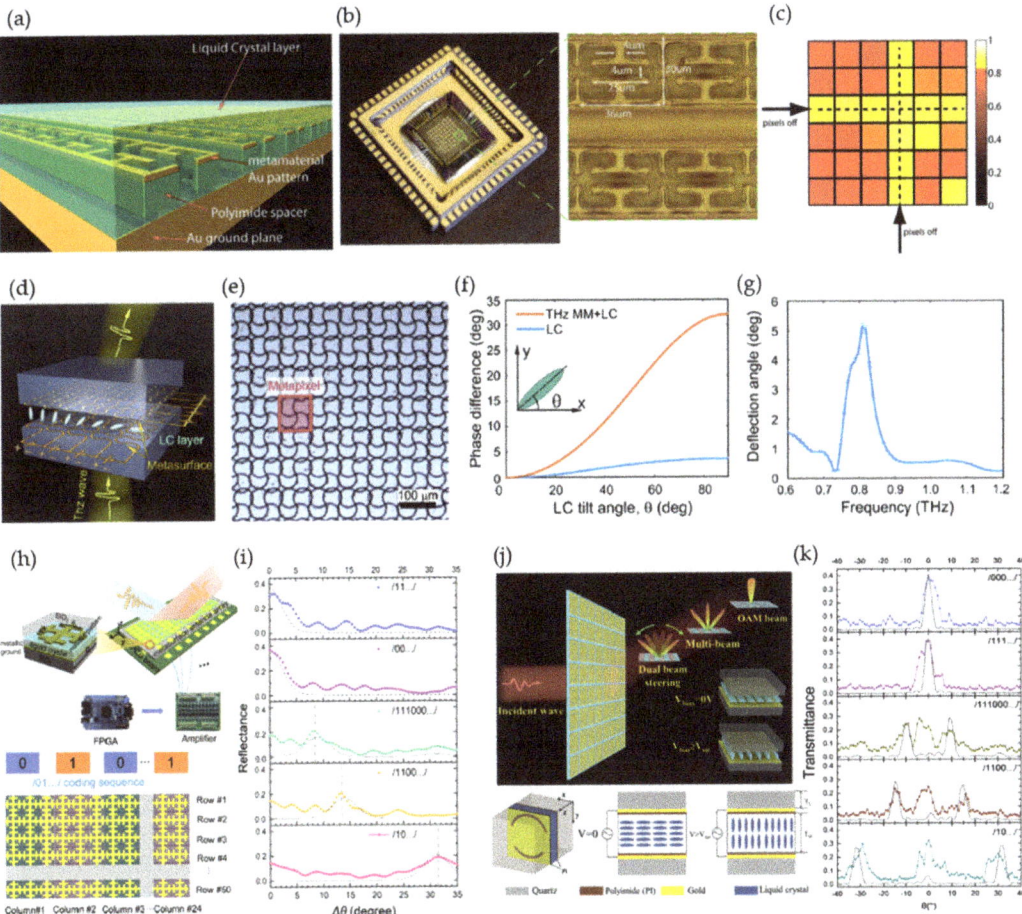

Figure 9. Liquid-crystal-enabled reconfigurable metasurface. (**a–c**) A spatial light modulator based on liquid crystals. (**a**) Schematic of metamaterial absorbers covered with a layer of liquid crystals. (**b**) Spatial light modulator device and an enlargement for the meta-atom dimensions. (**c**) A 6 × 6 pixelated absorption map measured at 3.725 THz. Reprinted from Ref. [102]. (**d–g**) A spatial phase modulator operating at 0.8 THz. (**d**) Schematic of the metasurface. (**e**) An optical microscopic image of the fabricated metasurface. (**f**) Phase difference as a function of liquid-crystal tilt angle. (**g**) Calculated beam deflection. Reprinted from Ref. [104]. (**h**,**i**) Programmable metasurface for beam steering. (**h**) Schematic of the beam steering metasurface with the control element (top). The unit cell consists of a liquid-crystal layer embedded between two metallic layers. Schematic of the metasurface with the applied coding sequence of /01.../(bottom). (**i**) Reflected angles for five different coding sequences with an incident angle of 20° at 0.672 THz. Reprinted from Ref. [105]. (**j**,**k**) Liquid crystal-based multifunctional transmissive coding metasurface. (**j**) Schematic of the functional metasurface (top) and the asymmetric unit-cell design (bottom). (**k**) Measured transmitted pattern for different coding sequences. Reprinted from Ref. [106].

The reconfigurable effective refractive index of liquid crystal makes it suitable for both amplitude and phase modulation [103]. Figure 9d shows a spatial phase modulator of a nematic liquid-crystal layer sandwiched between two orthogonally placed metasurfaces [104]. Meandering wires enabled the electric potential of each pixel to be selectively and spatially addressed. The anisotropic metapixel (unit cell) consisted of two metallic split rings to enhance liquid-crystal birefringence (Figure 9e,f). The maximum phase change (32°) resulted from a 90° tilt angle. A biasing voltage of 20 V achieved a deflection angle of 5° at 0.8 THz (Figure 9g). Combining liquid crystals with a programmable metasurface enabled more advanced dynamic beam steering (Figure 9h) [105]. A 25 µm thick liquid-crystal layer was embedded inside a metal–insulator–metal (MIM) resonator with a top metal layer patterned in the Jerusalem cross structure. Bias voltages of 0 and 40 V were applied to have "0" and "1" states with 180° phase differences while maintaining the same reflection amplitude. By changing the phase gradient through coding pattern, different reflected angles were achieved at 0.672 THz, for an incident angle of 20°, as shown in Figure 9i. The maximum acquired deflection angle was 32°, with a reflection efficiency of 19.1%. By designing unit cells with more phase changes, this method can be further extended to two-bit- or three-bit-coding liquid-crystal metasurfaces, achieving a wider beam-deflection angle and higher reflection efficiency. Recently, an liquid crystal -based transmissive coding metasurface was demonstrated for multifunctional control (Figure 9j) [106]. An asymmetric metasurface pattern was designed for Fano resonance, realizing a transmission efficiency as high as ~50% at 0.426 THz. Figure 9k shows the measured beam-splitting patterns using different coding sequences.

2.4. Micro-Electromechanical-System (MEMS)

In contrast to other tuning mechanisms that alter the properties of materials, micro-electromechanical-system (MEMS) metasurfaces directly change the structural geometry of the unit cell, transforming the electromagnetic wave responses. Moreover, advanced and developed MEMS manufacturing makes it attractive for reconfigurable THz devices [107]. The simplest microstructure cantilever has been extensively studied and was embedded in a metasurface for active tuning [44]. A wideband spatial light modulator was built using an array of 768 actuatable mirrors, with a length of 220 µm and a width of 100 µm (Figure 10a) [108]. These dimensions were selected to reduce diffraction from individual mirrors and to increase the pixel-to-pixel modulation contrast of the spatial light modulator. A cantilever consisting of chrome –copper–chrome multilayers had intrinsic residual stress forcing it to tilt up with an angle of 35° (Figure 10b), which minimized back-diffracted waves in the incident direction. The mirrors were pulled down to the substrate by applying a bias voltage of 37 V. The SLM was built with micromirror arrays based on the grating concept to have a wide operational spectral range. The authors designed a spatial light modulator with 4 × 6 independently switchable pixels. Each pixel consisted of 4 × 8 micromirrors with a pixel size of 1 mm × 2 mm (Figure 10c). The modulation contrast was higher than 50% over the frequency range from 0.97 THz to 2.28 THz, with a peak modulation contrast of 87% at 1.38 THz. This method allows for almost arbitrary spatial pixel sizes by collectively switching the corresponding group of mirrors.

Beam steering was demonstrated by a cantilever designed for electrical LC resonance (Figure 10d) [109]. By controlling the suspension angle (from 2° to 0°) of the bimorph cantilever through biasing voltage (from 0 V to 30 V), a phase coverage of 310° was achieved at 0.8 THz (Figure 10e). Steering angles of ±70° and ±39° were demonstrated through simulation by grouping twelve columns as a super cell and controlling each column individually (Figure 10f). Each unit cell could potentially be addressed separately to have unit-cell-level control to enable a programmable MEMS metasurface. A MEMS-based metasurface for wireless security encoding was demonstrated using square double SRRs, forming a Fano resonator (Figure 10g) [110]. Independent control of the two SRRs provided four transmission states for an exclusive-OR (XOR) logic operation (Figure 10h). An application performing the XOR logic operation for one-time-pad (OTP) security in wireless transmission is illustrated in Figure 10i. A private message (m) was encoded with a secret key (k) using the XOR logic operation before sending it out for public-channel transmission, and the original message was decrypted at the destination through the inverse XOR operation. This fast encryption method could be extended to other wireless communication networks requiring high security.

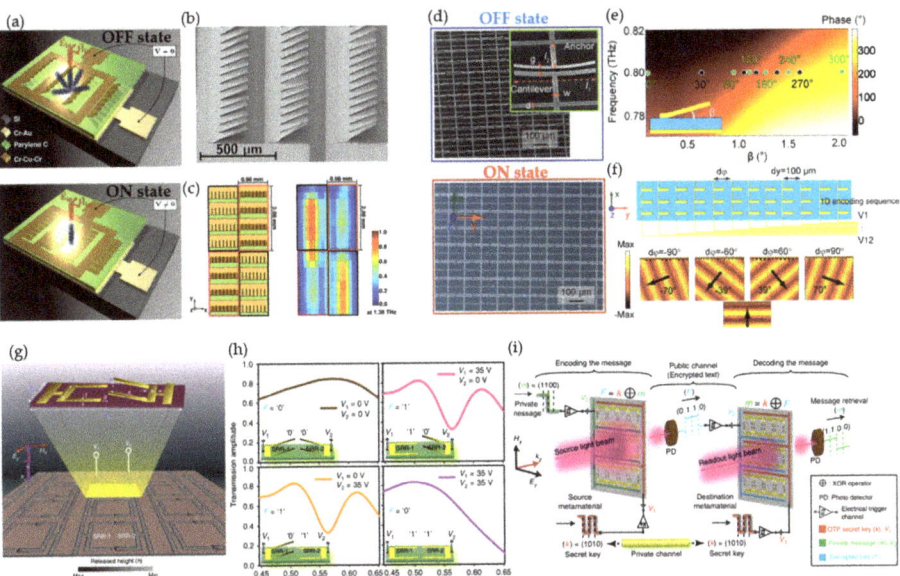

Figure 10. MEMS-enabled reconfigurable metasurface. (**a**–**c**) Micromirror array for the wideband spatial light modulator. (**a**) Schematic of a single-pixel in OFF state (top) and ON-state (bottom) for a bias voltage of 0 V and 37 V, respectively. (**b**) SEM image of the inclined mirrors. (**c**) Model of a 2 × 2-pixel SLM with two ON pixels (highlighted by black frames) along one diagonal (left) and its corresponding measured-intensity distribution (right) at 1.38 THz. Reprinted from Ref. [108]. (**d**–**f**) MEMS-based metal–insulator–metal metadevices for beam steering. (**d**) Images of the fabricated metasurface in "ON" and "OFF" states. (**e**) Simulated phase response as a function of the cantilever angles. (**f**) Simulated dynamic beam steering with six-digit control. Reprinted from Ref. [109]. (**g**–**i**) Reconfigurable MEMS Fano metasurfaces for logic operations in cryptographic wireless communication networks. (**g**) Unit-cell model of the metasurface. (**h**) Measured far-field transmission spectra showing the exclusive-OR (XOR) logic feature for various voltage states of the SRRs at 0.56 THz. (**i**) Implementation of the XOR logic for OTP-secured wireless communication channels. Reprinted from Ref. [110].

3. Discussion

THz RISs can dynamically modify the wave propagation direction, thereby enhancing wireless network efficiency, are crucial for actualization of 6G communication links. Reconfigurable metasurfaces with the function of pixel-level amplitude modulation and tunable beam steering are summarized in Tables 1 and 2, respectively, according to tuning elements. Electrical control allows for more precise, spatially selective pixel-level modulation but requires additional wires and complex control circuits. Optical control usually modulates the whole surface with a laser pump, but localization may be possible through an additional coding mask [101]. Tuning elements can be combined to add additional reconfigurable freedom [111,112]. Other reconfigurable methods, such as mechanical [113] and microfluid-based [114] tuning, are also feasible. RISs realized through industry-standard fabrication processes have greater potential for large-scale construction, which is required for future 6G smart cities. Moreover, emerging 2D van der Waals materials using surface-plasmon polaritons provide new approaches for dynamic tunning in the THz spectrum [115–120]. Graphene plasmons demonstrate lower loss than conventional metal-based plasmonic metasurfaces [121,122]. Recently discovered phonon polaritons in van der Waals crystals exhibit remarkably low losses [123–125]. These phonon polaritons can be tunned through chemical intercalation [126–128], twisted stacking [129], and heterostructures [130–133].

Table 1. Pixel-level amplitude-modulation performance comparison of various tuning elements for THz RIS.

Tuning Element	Tuning Mechanism	Control Method	Frequency (THz)	Array Size	No. of State/Pixel	Modulation Depth	Modulation Speed	Power Consumption	Ref.
CMOS transistor	Field-effect-transistor (FET)-based switch	Bias voltage	0.3	12 × 12	256	25 dB	5 GHz	~mW (1.2 V)	Exp. [52]
Schottky diode	Depletion of substrate charge carrier	Bias voltage Bias voltage	0.36 2.72, 3.27, 3.81, 4.34	4 × 4 8 × 8	2 2	40% 62%	3 kHz 12 MHz	14 V −26.5 V	Exp. [55] Exp. [56]
HEMT	Depletion of channel-carrier density	Bias voltage	0.45	2 × 2	2	36%	10 MHz	<1 mW (1 V)	Exp. [62]
Graphene	Fermi level	Bias voltage Bias voltage	0.59 0.1	4 × 4 16 × 16	2 2	~50 to ~30% 80%	6 kHz 1 kHz	−10 V 4 V	Exp. [71] Exp. [72]
GST	Joule heating	Bias current	0.69, 0.64, 0.6, 0.56	2 × 2	8	100%	15 s	850 mA	Exp. [98]
Liquid crystal	Birefringence effect	Bias voltage	3.670	6 × 6	2	75%	1 kHz	15 V	Exp. [102]
MEMS	Structural deformation	Bias voltage	0.97 to 2.28	4 × 6	2	50%	1 kHz	37 V	Exp. [108]

Table 2. Beam-steering performance comparison of various tuning elements for THz RIS.

Tuning Element	Tuning Mechanism	Control Method	Frequency (THz)	Deflection Angle	Modulation Speed	Power Consumption	Ref.
CMOS transistor	Field-effect-transistor (FET)-based switch	Bias voltage	0.3	±30°	5 GHz	240 µW (1.2 V)	Exp. [52]
Schottky diode	Depletion of substrate charge carrier	Bias voltage Bias voltage	0.4 0.55–0.83	36.1° 59.09°–34.88°	1 kHz 3 kHz	−13 V −10 V	Exp. [57] Exp. [58]
Graphene	Fermi level	Bias voltage Bias voltage Bias voltage	0.8–1.4 1.05 1	42°–23° 5°, 11°, 17°, 23° −5°, 10°, 17.5°,	ps ps 60 GHz	— — 26 V, −44 V	Sim. [73] Sim. [82] Exp. [83]
Silicon	Photoconductivity	Laser pulses Laser pulses	0.6–1 0.586	51°–28° ±34.7°	30 ps 14 ps	1.9 mJ/cm^2 5.0 mW	Exp. [84] Exp. [85]
VO$_2$	Phase change	Joule heating Joule heating	0.7–1 0.1	±28° ±22°, −14°, 12°	<1 kHz <1 kHz	— 10 mV	Sim. [90] Exp. [89]
GeTe	Phase change	Laser pulses	0.3	±30°	35 ns	190 mJ/cm^2	Exp. [101]
Liquid crystals	Birefringence effect	Bias voltage Bias voltage Bias voltage	0.8 0.672 0.426	±4.5° 8.5°, 13.5°, 31.5° −9°, −15°, −29°, 9°, 16°, 30°	75 ms 100 Hz 1 kHz	20 V 40 V 10 V	Exp. [104] Exp. [105] Exp. [106]
MEMS	Structural deformation	Bias voltage	0.8	±39°, ±70°	sub-MHz	20 V	Sim. [109]

Since the emergence of multiple-input, multiple-output (MIMO) technology for wireless communication, polarization modulation (PoM) has gained more and more attention thanks to its excellent signal distinguishability, thereby increasing spectrum efficiency [134,135]. Metasurfaces with active elements for polarization modulation at THz frequencies were investigated in [136–139] and were found to have great potential to be utilized for future 6G wireless communications. Another promising area for future THz RIS applications is space-time-coding digital metasurfaces with have multifrequency control for beam shaping and steering [140]. Cell-free massive MIMO for mobile access is expected in 6G based on massive MIMO in 5G. Hybrid RISs with MIMO constitute an important future research direction.

Author Contributions: Writing—original draft preparation, F.Y.; writing—review and editing, P.P. and N.W. All authors have read and agreed to the published version of the manuscript.

Funding: This work was supported by the Science and Engineering Research Council of A*STAR (Agency for Science, Technology and Research), Singapore, under Grant no (A18A5b0056) and the project titled "Terahertz Nanogap Metasurface for enabling 6G communications".

Conflicts of Interest: The authors declare no conflict of interest.

References

1. Yang, P.; Xiao, Y.; Xiao, M.; Li, S. 6G Wireless Communications: Vision and Potential Techniques. *IEEE Netw.* **2019**, *33*, 70–75. [CrossRef]
2. Giordani, M.; Polese, M.; Mezzavilla, M.; Rangan, S.; Zorzi, M. Toward 6G Networks: Use Cases and Technologies. *IEEE Commun. Mag.* **2020**, *58*, 55–61. [CrossRef]
3. Dang, S.; Amin, O.; Shihada, B.; Alouini, M.S. What Should 6G Be? *Nat. Electron.* **2020**, *3*, 20–29. [CrossRef]
4. Imoize, A.L.; Adedeji, O.; Tandiya, N.; Shetty, S. 6G Enabled Smart Infrastructure for Sustainable Society: Opportunities, Challenges, and Research Roadmap. *Sensors* **2021**, *21*, 1709. [CrossRef] [PubMed]
5. Akyildiz, I.F.; Kak, A.; Nie, S. 6G and Beyond: The Future of Wireless Communications Systems. *IEEE Access* **2020**, *8*, 133995–134030. [CrossRef]
6. Chowdhury, M.Z.; Shahjalal, M.; Ahmed, S.; Jang, Y.M. 6G Wireless Communication Systems: Applications, Requirements, Technologies, Challenges, and Research Directions. *IEEE Open J. Commun. Soc.* **2020**, *1*, 957–975. [CrossRef]
7. Kleine-Ostmann, T.; Nagatsuma, T. A Review on Terahertz Communications Research. *J. Infrared Millim. Terahertz Waves* **2011**, *32*, 143–171. [CrossRef]
8. Elayan, H.; Amin, O.; Shihada, B.; Shubair, R.M.; Alouini, M.-S. Terahertz Band: The Last Piece of RF Spectrum Puzzle for Communication Systems. *IEEE Open J. Commun. Soc.* **2019**, *1*, 1–32. [CrossRef]
9. O'Hara, J.F.; Ekin, S.; Choi, W.; Song, I. A Perspective on Terahertz Next-Generation Wireless Communications. *Technologies* **2019**, *7*, 43. [CrossRef]
10. Jamshed, M.A.; Nauman, A.; Abbasi, M.A.B.; Kim, S.W. Antenna Selection and Designing for THz Applications: Suitability and Performance Evaluation: A Survey. *IEEE Access* **2020**, *8*, 113246–113261. [CrossRef]
11. Peng, B.; Guan, K.; Rey, S.; Kurner, T. Power-Angular Spectra Correlation Based Two Step Angle of Arrival Estimation for Future Indoor Terahertz Communications. *IEEE Trans. Antennas Propag.* **2019**, *67*, 7097–7105. [CrossRef]
12. Liaskos, C.; Nie, S.; Tsioliaridou, A.; Pitsillides, A.; Ioannidis, S.; Akyildiz, I. A New Wireless Communication Paradigm through Software-Controlled Metasurfaces. *IEEE Commun. Mag.* **2018**, *56*, 162–169. [CrossRef]
13. Huang, C.; Zappone, A.; Alexandropoulos, G.C.; Debbah, M.; Yuen, C. Reconfigurable Intelligent Surfaces for Energy Efficiency in Wireless Communication. In *Proceedings of the IEEE Transactions on Wireless Communications*; Institute of Electrical and Electronics Engineers Inc.: Piscataway, NJ, USA, 2019; Volume 18, pp. 4157–4170.
14. Basar, E.; Renzo, M.D.; de Rosny, J.; Debbah, M.; Alouini, M.-S.; Zhang, R.; di Renzo, M. Wireless Communications Through Reconfigurable Intelligent Surfaces. *IEEE Access* **2019**, *7*, 116753–116773. [CrossRef]
15. Elmossallamy, M.A.; Zhang, H.; Song, L.; Seddik, K.G.; Han, Z.; Li, G.Y. Reconfigurable Intelligent Surfaces for Wireless Communications: Principles, Challenges, and Opportunities. *IEEE Trans. Cogn. Commun. Netw.* **2020**, *6*, 990–1002. [CrossRef]
16. Alexandropoulos, G.C.; Lerosey, G.; Debbah, M.; Fink, M. Reconfigurable Intelligent Surfaces and Metamaterials: The Potential of Wave Propagation Control for 6G Wireless Communications. *arXiv* **2020**, arXiv:2006.11346.
17. Di, B.; Zhang, H.; Song, L.; Li, Y.; Han, Z.; Poor, H.V. Hybrid Beamforming for Reconfigurable Intelligent Surface Based Multi-User Communications: Achievable Rates with Limited Discrete Phase Shifts. *IEEE J. Sel. Areas Commun.* **2020**, *38*, 1809–1822. [CrossRef]
18. Yang, B.; Cao, X.; Huang, C.; Guan, Y.L.; Yuen, C.; di Renzo, M.; Niyato, D.; Debbah, M.; Hanzo, L. Spectrum Learning-Aided Reconfigurable Intelligent Surfaces for "Green" 6G Networks. *IEEE Netw.* **2021**, *35*, 20–26. [CrossRef]

19. Liaskos, C.; Tsioliaridou, A.; Pitsillides, A.; Akyildiz, I.F.; Kantartzis, N.V.; Lalas, A.X.; Dimitropoulos, X.; Ioannidis, S.; Kafesaki, M.; Soukoulis, C.M. Design and Development of Software Defined Metamaterials for Nanonetworks. *IEEE Circuits Syst. Mag.* **2015**, *15*, 12–25. [CrossRef]
20. Abadal, S.; Liaskos, C.; Tsioliaridou, A.; Ioannidis, S.; Pitsillides, A.; Sole-Pareta, J.; Alarcon, E.; Cabellos-Aparicio, A. Computing and Communications for the Software-Defined Metamaterial Paradigm: A Context Analysis. *IEEE Access* **2017**, *5*, 6225–6235. [CrossRef]
21. Zhao, J. A Survey of Intelligent Reflecting Surfaces (IRSs): Towards 6G Wireless Communication Networks. *arXiv* **2019**, arXiv:1907.04789.
22. Pillay, N.; Xu, H. Large Intelligent Surfaces: Random Waypoint Mobility and Two-Way Relaying. *Int. J. Commun. Syst.* **2020**, *33*, e4505. [CrossRef]
23. Yu, N.; Capasso, F. Flat Optics with Designer Metasurfaces. *Nat. Mater.* **2014**, *13*, 139–150. [CrossRef] [PubMed]
24. Yu, N.; Genevet, P.; Kats, M.A.; Aieta, F.; Tetienne, J.-P.; Capasso, F.; Gaburro, Z. Light Propagation with Phase Discontinuities: Generalized Laws of Reflection and Refraction. *Science* **2011**, *334*, 333–337. [CrossRef] [PubMed]
25. Wong, J.P.S.; Epstein, A.; Eleftheriades, G.V. Reflectionless Wide-Angle Refracting Metasurfaces. *IEEE Antennas Wirel. Propag. Lett.* **2016**, *15*, 1293–1296. [CrossRef]
26. Chen, M.; Abdo-Sánchez, E.; Epstein, A.; Eleftheriades, G.V. Theory, Design, and Experimental Verification of a Reflectionless Bianisotropic Huygens' Metasurface for Wide-Angle Refraction. *Phys. Rev. B* **2018**, *97*, 125433. [CrossRef]
27. Ho, J.S.; Qiu, B.; Tanabe, Y.; Yeh, A.J.; Fan, S.; Poon, A.S.Y. Planar Immersion Lens with Metasurfaces. *Phys. Rev. B-Condens. Matter Mater. Phys.* **2015**, *91*, 1–8. [CrossRef]
28. Zhuang, Z.P.; Chen, R.; Fan, Z.B.; Pang, X.N.; Dong, J.W. High Focusing Efficiency in Subdiffraction Focusing Metalens. *Nanophotonics* **2019**, *8*, 1279–1289. [CrossRef]
29. Yang, F.; Raeker, B.O.; Nguyen, D.T.; Miller, J.D.; Xiong, Z.; Grbic, A.; Ho, J.S. Antireflection and Wavefront Manipulation with Cascaded Metasurfaces. *Phys. Rev. Appl.* **2020**, *14*, 064044. [CrossRef]
30. Cui, T.J.; Qi, M.Q.; Wan, X.; Zhao, J.; Cheng, Q. Coding Metamaterials, Digital Metamaterials and Programmable Metamaterials. *Light Sci. Appl.* **2014**, *3*, e218. [CrossRef]
31. della Giovampaola, C.; Engheta, N. Digital Metamaterials. *Nat. Mater.* **2014**, *13*, 1115–1121. [CrossRef]
32. Gao, L.H.; Cheng, Q.; Yang, J.; Ma, S.J.; Zhao, J.; Liu, S.; Chen, H.B.; He, Q.; Jiang, W.X.; Ma, H.F.; et al. Broadband Diffusion of Terahertz Waves by Multi-Bit Coding Metasurfaces. *Light Sci. Appl.* **2015**, *4*, e324. [CrossRef]
33. Liu, S.; Cui, T.J.; Zhang, L.; Xu, Q.; Wang, Q.; Wan, X.; Gu, J.Q.; Tang, W.X.; Qing Qi, M.; Han, J.G.; et al. Convolution Operations on Coding Metasurface to Reach Flexible and Continuous Controls of Terahertz Beams. *Adv. Sci.* **2016**, *3*, 1600156. [CrossRef] [PubMed]
34. Ma, Q.; Shi, C.B.; Bai, G.D.; Chen, T.Y.; Noor, A.; Cui, T.J. Beam-Editing Coding Metasurfaces Based on Polarization Bit and Orbital-Angular-Momentum-Mode Bit. *Adv. Opt. Mater.* **2017**, *5*, 1700548. [CrossRef]
35. Liu, S.; Cui, T.J. Concepts, Working Principles, and Applications of Coding and Programmable Metamaterials. *Adv. Opt. Mater.* **2017**, *5*, 1700624. [CrossRef]
36. Ma, Q.; Chen, L.; Jing, H.B.; Hong, Q.R.; Cui, H.Y.; Liu, Y.; Li, L.; Cui, T.J. Controllable and Programmable Nonreciprocity Based on Detachable Digital Coding Metasurface. *Adv. Opt. Mater.* **2019**, *7*, 1901285. [CrossRef]
37. Wu, L.W.; Ma, H.F.; Wu, R.Y.; Xiao, Q.; Gou, Y.; Wang, M.; Wang, Z.X.; Bao, L.; Wang, H.L.; Qing, Y.M.; et al. Transmission-Reflection Controls and Polarization Controls of Electromagnetic Holograms by a Reconfigurable Anisotropic Digital Coding Metasurface. *Adv. Opt. Mater.* **2020**, *8*, 2001065. [CrossRef]
38. Wan, X.; Qi, M.Q.; Chen, T.Y.; Cui, T.J. Field-Programmable Beam Reconfiguring Based on Digitally-Controlled Coding Metasurface. *Sci. Rep.* **2016**, *6*, 20663. [CrossRef]
39. Yang, H.; Cao, X.; Yang, F.; Gao, J.; Xu, S.; Li, M.; Chen, X.; Zhao, Y.; Zheng, Y.; Li, S. A Programmable Metasurface with Dynamic Polarization, Scattering and Focusing Control. *Sci. Rep.* **2016**, *6*, 35692. [CrossRef]
40. Han, R.; Hu, Z.; Wang, C.; Holloway, J.; Yi, X.; Kim, M.; Mawdsley, J. Filling the Gap: Silicon Terahertz Integrated Circuits Offer Our Best Bet. *IEEE Microw. Mag.* **2019**, *20*, 80–93. [CrossRef]
41. Abadal, S.; Cui, T.J.; Low, T.; Georgiou, J. Programmable Metamaterials for Software-Defined Electromagnetic Control: Circuits, Systems, and Architectures. *IEEE J. Emerg. Sel. Top. Circuits Syst.* **2020**, *10*, 6–19. [CrossRef]
42. Bao, L.; Cui, T.J. Tunable, Reconfigurable, and Programmable Metamaterials. *Microw. Opt. Technol. Lett.* **2020**, *62*, 9–32. [CrossRef]
43. Tsilipakos, O.; Tasolamprou, A.C.; Pitilakis, A.; Liu, F.; Wang, X.; Mirmoosa, M.S.; Tzarouchis, D.C.; Abadal, S.; Taghvaee, H.; Liaskos, C.; et al. Toward Intelligent Metasurfaces: The Progress from Globally Tunable Metasurfaces to Software-Defined Metasurfaces with an Embedded Network of Controllers. *Adv. Opt. Mater.* **2020**, *8*, 2000783. [CrossRef]
44. Pitchappa, P.; Kumar, A.; Singh, R.; Lee, C.; Wang, N. Terahertz MEMS Metadevices. *J. Micromech. Microeng.* **2021**, *31*, 113001. [CrossRef]
45. Xu, J.; Yang, R.; Fan, Y.; Fu, Q.; Zhang, F. A Review of Tunable Electromagnetic Metamaterials with Anisotropic Liquid Crystals. *Front. Phys.* **2021**, *9*, 67. [CrossRef]
46. Mandal, A.; Cui, Y.; McRae, L.; Gholipour, B. Reconfigurable Chalcogenide Phase Change Metamaterials: A Material, Device, and Fabrication Perspective. *J. Phys. Photonics* **2021**, *3*, 022005. [CrossRef]

47. Guo, T.; Argyropoulos, C. Recent Advances in Terahertz Photonic Technologies Based on Graphene and Their Applications. *Adv. Photonics Res.* **2021**, *2*, 2000168. [CrossRef]
48. Li, L.; Jun Cui, T.; Ji, W.; Liu, S.; Ding, J.; Wan, X.; Bo Li, Y.; Jiang, M.; Qiu, C.W.; Zhang, S. Electromagnetic Reprogrammable Coding-Metasurface Holograms. *Nat. Commun.* **2017**, *8*, 1–7. [CrossRef] [PubMed]
49. Zhang, X.G.; Jiang, W.X.; Jiang, H.L.; Wang, Q.; Tian, H.W.; Bai, L.; Luo, Z.J.; Sun, S.; Luo, Y.; Qiu, C.W.; et al. An Optically Driven Digital Metasurface for Programming Electromagnetic Functions. *Nat. Electron.* **2020**, *3*, 165–171. [CrossRef]
50. Huang, C.; Zhang, C.; Yang, J.; Sun, B.; Zhao, B.; Luo, X. Reconfigurable Metasurface for Multifunctional Control of Electromagnetic Waves. *Adv. Opt. Mater.* **2017**, *5*, 1700485. [CrossRef]
51. Luo, Z.; Wang, Q.; Zhang, X.G.; Wu, J.W.; Dai, J.Y.; Zhang, L.; Wu, H.T.; Zhang, H.C.; Ma, H.F.; Cheng, Q.; et al. Intensity-Dependent Metasurface with Digitally Reconfigurable Distribution of Nonlinearity. *Adv. Opt. Mater.* **2019**, *7*, 1900792. [CrossRef]
52. Venkatesh, S.; Lu, X.; Saeidi, H.; Sengupta, K. A High-Speed Programmable and Scalable Terahertz Holographic Metasurface Based on Tiled CMOS Chips. *Nat. Electron.* **2020**, *3*, 785–793. [CrossRef]
53. Liu, Y.; Sun, T.; Xu, Y.; Wu, X.; Bai, Z.; Sun, Y.; Li, H.; Zhang, H.; Chen, K.; Ruan, C.; et al. Active Tunable THz Metamaterial Array Implemented in CMOS Technology. *J. Phys. D Appl. Phys.* **2021**, *54*, 085107. [CrossRef]
54. Chen, H.T.; Padilla, W.J.; Zide, J.M.O.; Gossard, A.C.; Taylor, A.J.; Averitt, R.D. Active Terahertz Metamaterial Devices. *Nature* **2006**, *444*, 597–600. [CrossRef] [PubMed]
55. Chan, W.L.; Chen, H.T.; Taylor, A.J.; Brener, I.; Cich, M.J.; Mittleman, D.M. A Spatial Light Modulator for Terahertz Beams. *Appl. Phys. Lett.* **2009**, *94*, 213511. [CrossRef]
56. Shrekenhamer, D.; Montoya, J.; Krishna, S.; Padilla, W.J. Four-Color Metamaterial Absorber THz Spatial Light Modulator. *Adv. Opt. Mater.* **2013**, *1*, 905–909. [CrossRef]
57. Karl, N.; Reichel, K.; Chen, H.T.; Taylor, A.J.; Brener, I.; Benz, A.; Reno, J.L.; Mendis, R.; Mittleman, D.M. An Electrically Driven Terahertz Metamaterial Diffractive Modulator with More than 20 DB of Dynamic Range. *Appl. Phys. Lett.* **2014**, *104*, 091115. [CrossRef]
58. Su, X.; Ouyang, C.; Xu, N.; Cao, W.; Wei, X.; Song, G.; Gu, J.; Tian, Z.; O'Hara, J.F.; Han, J.; et al. Active Metasurface Terahertz Deflector with Phase Discontinuities. *Opt. Express* **2015**, *23*, 27152. [CrossRef]
59. Dyakonov, M.; Shur, M. Shallow Water Analogy for a Ballistic Field Effect Transistor: New Mechanism of Plasma Wave Generation by Dc Current. *Phys. Rev. Lett.* **1993**, *71*, 2465. [CrossRef]
60. Dyakonov, M.; Shur, M. Detection, Mixing, and Frequency Multiplication of Terahertz Radiation by Two-Dimensional Electronic Fluid. *IEEE Trans. Electron Devices* **1996**, *43*, 380–387. [CrossRef]
61. Shrekenhamer, D.; Rout, S.; Strikwerda, A.C.; Bingham, C.; Averitt, R.D.; Sonkusale, S.; Padilla, W.J. High Speed Terahertz Modulation from Metamaterials with Embedded High Electron Mobility Transistors. *Opt. Express* **2011**, *19*, 9968–9975. [CrossRef]
62. Rout, S.; Sonkusale, S.R. A Low-Voltage High-Speed Terahertz Spatial Light Modulator Using Active Metamaterial. *APL Photonics* **2016**, *1*, 086102. [CrossRef]
63. Nouman, M.T.; Kim, H.W.; Woo, J.M.; Hwang, J.H.; Kim, D.; Jang, J.H. Terahertz Modulator Based on Metamaterials Integrated with Metal-Semiconductor-Metal Varactors. *Sci. Rep.* **2016**, *6*, 26452. [CrossRef]
64. Zhao, Y.; Wang, L.; Zhang, Y.; Qiao, S.; Liang, S.; Zhou, T.; Zhang, X.; Guo, X.; Feng, Z.; Lan, F.; et al. High-Speed Efficient Terahertz Modulation Based on Tunable Collective-Individual State Conversion within an Active 3 Nm Two-Dimensional Electron Gas Metasurface. *Nano Lett.* **2019**, *19*, 7588–7597. [CrossRef]
65. Lee, G.; Nouman, M.T.; Hwang, J.H.; Kim, H.W.; Jang, J.H. Enhancing the Modulation Depth of a Dynamic Terahertz Metasurface by Integrating into an Asymmetric Fabry-Pérot Cavity. *AIP Adv.* **2018**, *8*, 095310. [CrossRef]
66. Zhang, Y.; Qiao, S.; Liang, S.; Wu, Z.; Yang, Z.; Feng, Z.; Sun, H.; Zhou, Y.; Sun, L.; Chen, Z.; et al. Gbps Terahertz External Modulator Based on a Composite Metamaterial with a Double-Channel Heterostructure. *Nano Lett.* **2015**, *15*, 3501–3506. [CrossRef]
67. Zhang, Y.; Zhao, Y.; Liang, S.; Zhang, B.; Wang, L.; Zhou, T.; Kou, W.; Lan, F.; Zeng, H.; Han, J.; et al. Large Phase Modulation of THz Wave via an Enhanced Resonant Active HEMT Metasurface. *Nanophotonics* **2018**, *8*, 153–170. [CrossRef]
68. Carrasco, E.; Tamagnone, M.; Perruisseau-Carrier, J. Tunable Graphene Reflective Cells for THz Reflectarrays and Generalized Law of Reflection. *Appl. Phys. Lett.* **2013**, *102*, 104103. [CrossRef]
69. Wang, R.; Ren, X.G.; Yan, Z.; Jiang, L.J.; Sha, W.E.I.; Shan, G.C. Graphene Based Functional Devices: A Short Review. *Front. Phys.* **2019**, *14*, 13603. [CrossRef]
70. Sensale-Rodriguez, B.; Yan, R.; Rafique, S.; Zhu, M.; Li, W.; Liang, X.; Gundlach, D.; Protasenko, V.; Kelly, M.M.; Jena, D.; et al. Extraordinary Control of Terahertz Beam Reflectance in Graphene Electro-Absorption Modulators. *Nano Lett.* **2012**, *12*, 4518–4522. [CrossRef] [PubMed]
71. Sensale-Rodriguez, B.; Rafique, S.; Yan, R.; Zhu, M.; Protasenko, V.; Jena, D.; Liu, L.; Xing, H.G. Terahertz Imaging Employing Graphene Modulator Arrays. *Opt. Express* **2013**, *21*, 2324–2330. [CrossRef]
72. Malevich, Y.; Ergoktas, M.S.; Bakan, G.; Steiner, P.; Kocabas, C. Video-Speed Graphene Modulator Arrays for Terahertz Imaging Applications. *ACS Photonics* **2020**, *7*, 2374–2380. [CrossRef]
73. Chen, D.; Yang, J.; Huang, J.; Bai, W.; Zhang, J.; Zhang, Z.; Xu, S.; Xie, W. The Novel Graphene Metasurfaces Based on Split-Ring Resonators for Tunable Polarization Switching and Beam Steering at Terahertz Frequencies. *Carbon* **2019**, *154*, 350–356. [CrossRef]

74. Chen, D.; Yang, J.; Huang, J.; Zhang, Z.; Xie, W.; Jiang, X.; He, X.; Han, Y.; Zhang, Z.; Yu, Y. Continuously Tunable Metasurfaces Controlled by Single Electrode Uniform Bias-Voltage Based on Nonuniform Periodic Rectangular Graphene Arrays. *Opt. Express* **2020**, *28*, 29306. [CrossRef] [PubMed]
75. Zhang, Y.; Feng, Y.; Zhao, J.; Jiang, T.; Zhu, B. Terahertz Beam Switching by Electrical Control of Graphene-Enabled Tunable Metasurface. *Sci. Rep.* **2017**, *7*, 14147. [CrossRef]
76. Xu, J.; Liu, W.; Song, Z. Graphene-Based Terahertz Metamirror with Wavefront Reconfiguration. *Opt. Express* **2021**, *29*, 39574. [CrossRef]
77. Xiao, B.; Zhang, Y.; Tong, S.; Yu, J.; Xiao, L. Novel Tunable Graphene-Encoded Metasurfaces on an Uneven Substrate for Beam-Steering in Far-Field at the Terahertz Frequencies. *Opt. Express* **2020**, *28*, 7125. [CrossRef]
78. Xu, J.; Liu, W.; Song, Z. Terahertz Dynamic Beam Steering Based on Graphene Coding Metasurfaces. *IEEE Photonics J.* **2021**, *13*, 4600409. [CrossRef]
79. Momeni, A.; Rouhi, K.; Rajabalipanah, H.; Abdolali, A. An Information Theory-Inspired Strategy for Design of Re-Programmable Encrypted Graphene-Based Coding Metasurfaces at Terahertz Frequencies. *Sci. Rep.* **2018**, *8*, 6200. [CrossRef]
80. Hosseininejad, S.E.; Rouhi, K.; Neshat, M.; Faraji-Dana, R.; Cabellos-Aparicio, A.; Abadal, S.; Alarcón, E. Reprogramming Graphene-Based Metasurface Mirror with Adaptive Focal Point for THz Imaging. *Sci. Rep.* **2019**, *9*, 2868. [CrossRef]
81. Hosseininejad, S.E.; Rouhi, K.; Neshat, M.; Cabellos-Aparicio, A.; Abadal, S.; Alarcon, E. Digital Metasurface Based on Graphene: An Application to Beam Steering in Terahertz Plasmonic Antennas. *IEEE Trans. Nanotechnol.* **2019**, *18*, 734–746. [CrossRef]
82. Wang, B.; Luo, X.; Lu, Y.; Li, G. Full 360° Terahertz Dynamic Phase Modulation Based on Doubly Resonant Graphene–Metal Hybrid Metasurfaces. *Nanomaterials* **2021**, *11*, 3157. [CrossRef] [PubMed]
83. Tamagnone, M.; Capdevila, S.; Lombardo, A.; Wu, J.; Centeno, A.; Zurutuza, A.; Ionescu, A.M.; Ferrari, A.C.; Mosig, J.R. Graphene Reflectarray Metasurface for Terahertz Beam Steering and Phase Modulation. *arXiv* **2018**, arXiv:1806.02202.
84. Cong, L.; Srivastava, Y.K.; Zhang, H.; Zhang, X.; Han, J.; Singh, R. All-Optical Active THz Metasurfaces for Ultrafast Polarization Switching and Dynamic Beam Splitting. *Light Sci. Appl.* **2018**, *7*, 28. [CrossRef]
85. Cong, L.; Singh, R. Spatiotemporal Dielectric Metasurfaces for Unidirectional Propagation and Reconfigurable Steering of Terahertz Beams. *Adv. Mater.* **2020**, *32*, 2001418. [CrossRef]
86. Wuttig, M.; Bhaskaran, H.; Taubner, T. Phase-Change Materials for Non-Volatile Photonic Applications. *Nat. Photonics* **2017**, *11*, 465–476. [CrossRef]
87. Raeis-Hosseini, N.; Rho, J. Metasurfaces Based on Phase-Change Material as a Reconfigurable Platform for Multifunctional Devices. *Materials* **2017**, *10*, 1046. [CrossRef] [PubMed]
88. Wang, L.; Zhang, Y.; Guo, X.; Chen, T.; Liang, H.; Hao, X.; Hou, X.; Kou, W.; Zhao, Y.; Zhou, T.; et al. A Review of THz Modulators with Dynamic Tunable Metasurfaces. *Nanomaterials* **2019**, *9*, 965. [CrossRef]
89. Hashemi, M.R.M.; Yang, S.H.; Wang, T.; Sepúlveda, N.; Jarrahi, M. Electronically-Controlled Beam-Steering through Vanadium Dioxide Metasurfaces. *Sci. Rep.* **2016**, *6*, 35439. [CrossRef]
90. Ding, F.; Zhong, S.; Bozhevolnyi, S.I. Vanadium Dioxide Integrated Metasurfaces with Switchable Functionalities at Terahertz Frequencies. *Adv. Opt. Mater.* **2018**, *6*, 1701204. [CrossRef]
91. Li, J.; Yang, Y.; Li, J.; Zhang, Y.; Zhang, Z.; Zhao, H.; Li, F.; Tang, T.; Dai, H.; Yao, J. All-Optical Switchable Vanadium Dioxide Integrated Coding Metasurfaces for Wavefront and Polarization Manipulation of Terahertz Beams. *Adv. Theory Simul.* **2020**, *3*, 1900783. [CrossRef]
92. Wang, S.; Kang, L.; Werner, D.H. Hybrid Resonators and Highly Tunable Terahertz Metamaterials Enabled by Vanadium Dioxide (VO_2). *Sci. Rep.* **2017**, *7*, 4286. [CrossRef] [PubMed]
93. Wang, H.; Deng, L.; Zhang, C.; Qu, M.; Wang, L.; Li, S. Dual-Band Reconfigurable Coding Metasurfaces Hybridized with Vanadium Dioxide for Wavefront Manipulation at Terahertz Frequencies. *Microw. Opt. Technol. Lett.* **2019**, *61*, 2847–2853. [CrossRef]
94. Pan, W.-M.; Li, J.-S.; Zhou, C. Switchable Digital Metasurface Based on Phase Change Material in the Terahertz Region. *Opt. Mater. Express* **2021**, *11*, 1070. [CrossRef]
95. Shabanpour, J.; Beyraghi, S.; Cheldavi, A. Ultrafast Reprogrammable Multifunctional Vanadium-Dioxide-Assisted Metasurface for Dynamic THz Wavefront Engineering. *Sci. Rep.* **2020**, *10*, 1–14. [CrossRef]
96. Jiang, M.; Hu, F.; Zhang, L.; Quan, B.; Xu, W.; Du, H.; Xie, D.; Chen, Y. Electrically Triggered VO2 Reconfigurable Metasurface for Amplitude and Phase Modulation of Terahertz Wave. *J. Lightwave Technol.* **2021**, *39*, 3488–3494. [CrossRef]
97. Chen, B.; Wu, J.; Li, W.; Zhang, C.; Fan, K.; Xue, Q.; Chi, Y.; Wen, Q.; Jin, B.; Chen, J.; et al. Programmable Terahertz Metamaterials with Non-Volatile Memory. *Laser Photonics Reviews* **2022**, 2100472. [CrossRef]
98. Pitchappa, P.; Kumar, A.; Prakash, S.; Jani, H.; Venkatesan, T.; Singh, R. Chalcogenide Phase Change Material for Active Terahertz Photonics. *Adv. Mater.* **2019**, *31*, 1808157. [CrossRef] [PubMed]
99. Kodama, C.H.; Coutu, R.A. Tunable Split-Ring Resonators Using Germanium Telluride. *Appl. Phys. Lett.* **2016**, *108*, 231901. [CrossRef]
100. Gwin, A.H.; Kodama, C.H.; Laurvick, T.V.; Coutu, R.A.; Taday, P.F. Improved Terahertz Modulation Using Germanium Telluride (GeTe) Chalcogenide Thin Films. *Appl. Phys. Lett.* **2015**, *107*, 031904. [CrossRef]
101. Lin, Q.W.; Wong, H.; Huitema, L.; Crunteanu, A. Coding Metasurfaces with Reconfiguration Capabilities Based on Optical Activation of Phase-Change Materials for Terahertz Beam Manipulations. *Adv. Opt. Mater.* **2021**, *10*, 2101699. [CrossRef]

102. Savo, S.; Shrekenhamer, D.; Padilla, W.J. Liquid Crystal Metamaterial Absorber Spatial Light Modulator for THz Applications. *Adv. Opt. Mater.* **2014**, *2*, 275–279. [CrossRef]
103. Vasic, B.; Isic, G.; Beccherelli, R.; Zografopoulos, D.C. Tunable Beam Steering at Terahertz Frequencies Using Reconfigurable Metasurfaces Coupled with Liquid Crystals. *IEEE J. Sel. Top. Quantum Electron.* **2020**, *26*, 7701609. [CrossRef]
104. Buchnev, O.; Podoliak, N.; Kaltenecker, K.; Walther, M.; Fedotov, V.A. Metasurface-Based Optical Liquid Crystal Cell as an Ultrathin Spatial Phase Modulator for THz Applications. *ACS Photonics* **2020**, *7*, 3199–3206. [CrossRef]
105. Wu, J.; Shen, Z.; Ge, S.; Chen, B.; Shen, Z.; Wang, T.; Zhang, C.; Hu, W.; Fan, K.; Padilla, W.; et al. Liquid Crystal Programmable Metasurface for Terahertz Beam Steering. *Appl. Phys. Lett.* **2020**, *116*, 131104. [CrossRef]
106. Liu, C.X.; Yang, F.; Fu, X.J.; Wu, J.W.; Zhang, L.; Yang, J.; Cui, T.J. Programmable Manipulations of Terahertz Beams by Transmissive Digital Coding Metasurfaces Based on Liquid Crystals. *Adv. Opt. Mater.* **2021**, *9*, 2100932. [CrossRef]
107. Oberhammer, J. THz MEMS-Micromachining Enabling New Solutions at Millimeter and Submillimeter Frequencies. In Proceedings of the 2016 Global Symposium on Millimeter Waves, GSMM 2016 and ESA Workshop on Millimetre-Wave Technology and Applications, Kuala Lumpur, Malaysia, 13–16 November 2017.
108. Kappa, J.; Sokoluk, D.; Klingel, S.; Shemelya, C.; Oesterschulze, E.; Rahm, M. Electrically Reconfigurable Micromirror Array for Direct Spatial Light Modulation of Terahertz Waves over a Bandwidth Wider Than 1 THz. *Sci. Rep.* **2019**, *9*, 2597. [CrossRef]
109. Cong, L.; Pitchappa, P.; Wu, Y.; Ke, L.; Lee, C.; Singh, N.; Yang, H.; Singh, R. Active Multifunctional Microelectromechanical System Metadevices: Applications in Polarization Control, Wavefront Deflection, and Holograms. *Adv. Opt. Mater.* **2017**, *5*, 1600716. [CrossRef]
110. Manjappa, M.; Pitchappa, P.; Singh, N.; Wang, N.; Zheludev, N.I.; Lee, C.; Singh, R. Reconfigurable MEMS Fano Metasurfaces with Multiple-Input–Output States for Logic Operations at Terahertz Frequencies. *Nat. Commun.* **2018**, *9*, 4056. [CrossRef]
111. Shen, Y.; Wang, J.; Wang, Q.; Qiao, X.; Wang, Y.; Xu, D. Broadband Tunable Terahertz Beam Deflector Based on Liquid Crystals and Graphene. *Crystals* **2021**, *11*, 1141. [CrossRef]
112. Li, H.; Xu, W.; Cui, Q.; Wang, Y.; Yu, J. Theoretical Design of a Reconfigurable Broadband Integrated Metamaterial Terahertz Device. *Opt. Express* **2020**, *28*, 40060. [CrossRef]
113. Chen, K.; Zhang, X.; Chen, X.; Wu, T.; Wang, Q.; Zhang, Z.; Xu, Q.; Han, J.; Zhang, W. Active Dielectric Metasurfaces for Switchable Terahertz Beam Steering and Focusing. *IEEE Photonics J.* **2021**, *13*, 4600111. [CrossRef]
114. Zhang, W.; Zhang, B.; Fang, X.; Cheng, K.; Chen, W.; Wang, Z.; Hong, D.; Zhang, M. Microfluid-Based Soft Metasurface for Tunable Optical Activity in THz Wave. *Opt. Express* **2021**, *29*, 8786. [CrossRef]
115. Caldwell, J.D.; Lindsay, L.; Giannini, V.; Vurgaftman, I.; Reinecke, T.L.; Maier, S.A.; Glembocki, O.J. Low-Loss, Infrared and Terahertz Nanophotonics Using Surface Phonon Polaritons. *Nanophotonics* **2015**, *4*, 44–68. [CrossRef]
116. Basov, D.N.; Fogler, M.M.; García De Abajo, F.J. Polaritons in van Der Waals Materials. *Science* **2016**, *354*, aag1992. [CrossRef] [PubMed]
117. Low, T.; Chaves, A.; Caldwell, J.D.; Kumar, A.; Fang, N.X.; Avouris, P.; Heinz, T.F.; Guinea, F.; Martin-Moreno, L.; Koppens, F. Polaritons in Layered Two-Dimensional Materials. *Nat. Mater.* **2017**, *16*, 182–194. [CrossRef]
118. Dai, Z.; Hu, G.; Ou, Q.; Zhang, L.; Xia, F.; Garcia-Vidal, F.J.; Qiu, C.W.; Bao, Q. Artificial Metaphotonics Born Naturally in Two Dimensions. *Chem. Rev.* **2020**, *120*, 6197–6246. [CrossRef] [PubMed]
119. Ma, W.; Shabbir, B.; Ou, Q.; Dong, Y.; Chen, H.; Li, P.; Zhang, X.; Lu, Y.; Bao, Q. Anisotropic Polaritons in van Der Waals Materials. *InfoMat* **2020**, *2*, 777–790. [CrossRef]
120. Song, M.; Jayathurathnage, P.; Zanganeh, E.; Krasikova, M.; Smirnov, P.; Belov, P.; Kapitanova, P.; Simovski, C.; Tretyakov, S.; Krasnok, A. Wireless Power Transfer Based on Novel Physical Concepts. *Nat. Electron.* **2021**, *4*, 707–716. [CrossRef]
121. Ni, G.X.; McLeod, A.S.; Sun, Z.; Wang, L.; Xiong, L.; Post, K.W.; Sunku, S.S.; Jiang, B.Y.; Hone, J.; Dean, C.R.; et al. Fundamental Limits to Graphene Plasmonics. *Nature* **2018**, *557*, 530–533. [CrossRef]
122. Alonso-González, P.; Nikitin, A.Y.; Gao, Y.; Woessner, A.; Lundeberg, M.B.; Principi, A.; Forcellini, N.; Yan, W.; Vélez, S.; Huber, A.J.; et al. Acoustic Terahertz Graphene Plasmons Revealed by Photocurrent Nanoscopy. *Nat. Nanotechnol.* **2017**, *12*, 31–35. [CrossRef]
123. Walsh, B.M.; Foster, J.C.; Erickson, P.J.; Sibeck, D.G. Tunable Phonon Polaritons in Atomically Thin van Der Waals Crystals of Boron Nitride. *Science* **2014**, *343*, 1122–1125. [CrossRef]
124. Ma, W.; Alonso-González, P.; Li, S.; Nikitin, A.Y.; Yuan, J.; Martín-Sánchez, J.; Taboada-Gutiérrez, J.; Amenabar, I.; Li, P.; Vélez, S.; et al. In-Plane Anisotropic and Ultra-Low-Loss Polaritons in a Natural van Der Waals Crystal. *Nature* **2018**, *562*, 557–562. [CrossRef] [PubMed]
125. Zheng, Z.; Xu, N.; Oscurato, S.L.; Tamagnone, M.; Sun, F.; Jiang, Y.; Ke, Y.; Chen, J.; Huang, W.; Wilson, W.L.; et al. A Mid-Infrared Biaxial Hyperbolic van Der Waals Crystal. *Sci. Adv.* **2019**, *5*, eaav86902019. [CrossRef]
126. Wu, Y.; Ou, Q.; Yin, Y.; Li, Y.; Ma, W.; Yu, W.; Liu, G.; Cui, X.; Bao, X.; Duan, J.; et al. Chemical Switching of Low-Loss Phonon Polaritons in α-MoO3 by Hydrogen Intercalation. *Nat. Commun.* **2020**, *11*, 2646. [CrossRef]
127. Taboada-Gutiérrez, J.; Álvarez-Pérez, G.; Duan, J.; Ma, W.; Crowley, K.; Prieto, I.; Bylinkin, A.; Autore, M.; Volkova, H.; Kimura, K.; et al. Broad Spectral Tuning of Ultra-Low-Loss Polaritons in a van Der Waals Crystal by Intercalation. *Nat. Mater.* **2020**, *19*, 964–968. [CrossRef]
128. Wu, Y.; Ou, Q.; Dong, S.; Hu, G.; Si, G.; Dai, Z.; Qiu, C.W.; Fuhrer, M.S.; Mokkapati, S.; Bao, Q. Efficient and Tunable Reflection of Phonon Polaritons at Built-In Intercalation Interfaces. *Adv. Mater.* **2021**, *33*, 2008070. [CrossRef] [PubMed]

129. Hu, G.; Ou, Q.; Si, G.; Wu, Y.; Wu, J.; Dai, Z.; Krasnok, A.; Mazor, Y.; Zhang, Q.; Bao, Q.; et al. Topological Polaritons and Photonic Magic Angles in Twisted α-MoO3 Bilayers. *Nature* **2020**, *582*, 209–213. [CrossRef] [PubMed]
130. Dai, S.; Ma, Q.; Liu, M.K.; Andersen, T.; Fei, Z.; Goldflam, M.D.; Wagner, M.; Watanabe, K.; Taniguchi, T.; Thiemens, M.; et al. Graphene on Hexagonal Boron Nitride as a Tunable Hyperbolic Metamaterial. *Nat. Nanotechnol.* **2015**, *10*, 682–686. [CrossRef]
131. Zhang, Q.; Ou, Q.; Hu, G.; Liu, J.; Dai, Z.; Fuhrer, M.S.; Bao, Q.; Qiu, C.W. Hybridized Hyperbolic Surface Phonon Polaritons at α-MoO3 and Polar Dielectric Interfaces. *Nano Lett.* **2021**, *21*, 3112–3119. [CrossRef]
132. Álvarez-Pérez, G.; González-Morán, A.; Capote-Robayna, N.; Voronin, K.V.; Duan, J.; Volkov, V.S.; Alonso-González, P.; Nikitin, A.Y. Active Tuning of Highly Anisotropic Phonon Polaritons in Van Der Waals Crystal Slabs by Gated Graphene. *ACS Photonics* **2022**. [CrossRef]
133. Zeng, Y.; Ou, Q.; Liu, L.; Zheng, C.; Wang, Z.; Gong, Y.; Liang, X.; Zhang, Y.; Hu, G.; Yang, Z.; et al. Tailoring Topological Transition of Anisotropic Polaritons by Interface Engineering in Biaxial Crystals. *arXiv* **2022**, arXiv:2201.01412.
134. Huang, C.X.; Zhang, J.; Cheng, Q.; Cui, T.J. Polarization Modulation for Wireless Communications Based on Metasurfaces. *Adv. Funct. Mater.* **2021**, *31*, 2103379. [CrossRef]
135. Chen, X.; Ke, J.C.; Tang, W.; Chen, M.Z.; Dai, J.Y.; Basar, E.; Jin, S.; Cheng, Q.; Cui, T.J. Design and Implementation of MIMO Transmission Based on Dual-Polarized Reconfigurable Intelligent Surface. *IEEE Wirel. Commun. Lett.* **2021**, *10*, 2155–2159. [CrossRef]
136. Wong, H.; Wang, K.X.; Huitema, L.; Crunteanu, A. Active Meta Polarizer for Terahertz Frequencies. *Sci. Rep.* **2020**, *10*, 15382. [CrossRef] [PubMed]
137. Nakanishi, T.; Nakata, Y.; Urade, Y.; Okimura, K. Broadband Operation of Active Terahertz Quarter-Wave Plate Achieved with Vanadium-Dioxide-Based Metasurface Switchable by Current Injection. *Appl. Phys. Lett.* **2020**, *117*, 091102. [CrossRef]
138. Zhang, M.; Zhang, W.; Liu, A.Q.; Li, F.C.; Lan, C.F. Tunable Polarization Conversion and Rotation Based on a Reconfigurable Metasurface. *Sci. Rep.* **2017**, *7*, 091102. [CrossRef] [PubMed]
139. Lee, W.S.L.; Nirantar, S.; Headland, D.; Bhaskaran, M.; Sriram, S.; Fumeaux, C.; Withayachumnankul, W. Broadband Terahertz Circular-Polarization Beam Splitter. *Adv. Opt. Mater.* **2018**, *6*, 1700852. [CrossRef]
140. Castaldi, G.; Zhang, L.; Moccia, M.; Hathaway, A.Y.; Tang, W.X.; Cui, T.J.; Galdi, V. Joint Multi-Frequency Beam Shaping and Steering via Space–Time-Coding Digital Metasurfaces. *Adv. Funct. Mater.* **2021**, *31*, 2007620. [CrossRef]

Article
THz MEMS Switch Design

Yukang Feng *, Han-yu Tsao and N. Scott Barker *

Department of Electrical and Computer Engineering, University of Virginia, Charlottesville, VA 22904, USA; ht5fy@virginia.edu
* Correspondence: yf4rs@virginia.edu (Y.F.); nsb6t@virginia.edu (N.S.B.)

Abstract: In this work, an mm-wave/THz MEMS switch design process is presented. The challenges and solutions associated with the switch electrical design, modeling, fabrication, and test are explored and discussed. To investigate the feasibility of this design process, the switches are designed on both silicon and fused quartz substrate and then tested in the 140–750 GHz frequency range. The measurement fits design expectations and simulation well. At 750 GHz the measurement results from switches on both substrates have an ON state insertion loss of less than 3 dB and an OFF state isolation larger than 12 dB.

Keywords: millimeter-wave; terahertz; MEMS; switch; transmission line model

Citation: Feng, Y.; Tsao, H.-y.; Barker, N.S. THz MEMS Switch Design. *Micromachines* 2022, 13, 745. https://doi.org/10.3390/mi13050745

Academic Editors: Lu Zhang, Xiaodan Pang and Prakash Pitchappa

Received: 1 April 2022
Accepted: 4 May 2022
Published: 8 May 2022

Publisher's Note: MDPI stays neutral with regard to jurisdictional claims in published maps and institutional affiliations.

Copyright: © 2022 by the authors. Licensee MDPI, Basel, Switzerland. This article is an open access article distributed under the terms and conditions of the Creative Commons Attribution (CC BY) license (https://creativecommons.org/licenses/by/4.0/).

1. Introduction

Compared with conventional diode-based RF switches, MEMS switches have significant advantages in RF performance including higher isolation, lower insertion loss, and fewer intermodulation products. Meanwhile, since the MEMS switch does not require constant DC bias current in the static ON and OFF states, it consumes nearly zero power [1]. Because of these advantages, significant effort has been made to develop MEMS switches in the centimeter band [2–6], and some had also been successfully introduced in commercial applications [7,8]. In comparison, MEMS switch development in the millimeter-wave or THz spectrum faces more challenges.

In such frequencies, the switch's physical dimensions are comparable with its RF signal wavelength. Instead of regarding the switch as a lumped element, one needs to model the circuit from a transmission line perspective. The EM finite analysis is also frequently applied in RF optimization. Meanwhile, at such high frequencies, the switch RF performance improvement requires adjusting the circuit's physical features in dimensions of micrometers. Inevitably, the electrical design has to trade off with the fabrication techniques and limitations; which makes mm-wave/THz MEMS switch design more challenging.

In 2010, a DC contact MEMS switch operating at 50–100 GHz was reported [9]. Its center conductor in the coplanar waveguide (CPW) was actuated through a long cantilever beam from one side of the switch. The cantilever was driven by a comb-electrode actuated folded spring structure, which significantly complicated the bias structure and increased the circuit size. Another CMOS-based MEMS switch was successfully demonstrated at 220 GHz [10]. In this design, the air bridge structure that supported the actuator introduced a large parasitic capacitance and limited the isolation performance. Two switches had to be placed in series to provide 12 dB isolation. The switch reported in 2017 [11] successfully integrated a MEMS switch with a BiCMOS process. In the 220–320 GHz band this capacitive switch achieved 1 dB insertion loss and 12 dB isolation. Another 500–750 GHz waveguide switch [12] reported in 2017 used a MEMS-reconfigurable surface to block & unblock the wave propagation through the waveguide; however, this structure is incompatible with planar RF circuits.

In our previous work [13], we demonstrated an RF MEMS switch on a silicon substrate and provided the preliminary measurement in the 500–750 GHz (WR1.5) band. To investigate the feasibility of this design process on different dielectric substrates in a wider millimeter/THz spectrum, the switches were designed on both high resistivity silicon and fused quartz substrate and then tested in the 140–750 GHz frequency range. In this work, we provide a method to design RF MEMS switches in the mm-wave/THz frequency band. The challenges associated with the switch electrical design, modeling, fabrication, and test are discussed and the solutions are provided.

2. Mechanical Design

In this work, we focus on developing a series switch using a CPW structure as shown in Figure 1. Under applied external bias voltage, the electrostatic force pulls the cantilever towards the bias pad. The voltage that provides sufficient force that can turn the switch to an ON state is the actuation voltage. Once the external bias voltage reduces, the cantilever starts to restore to its initial position and returns to the OFF state under zero bias. In Figure 2, the switch's ON/OFF states are presented.

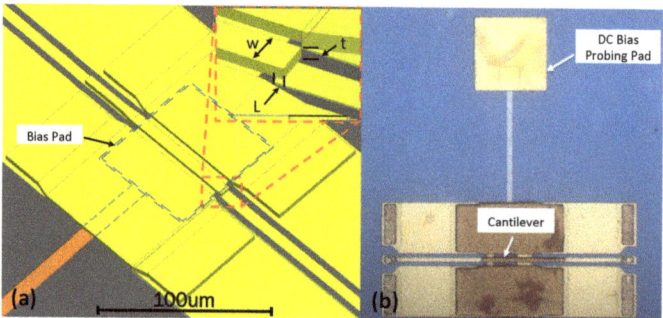

Figure 1. (a) 3-D model of an electro-static actuated MEMS switch in a series configuration. The switch's OFF state isolation is provided an air gap between the actuator's tip and the CPW section's tip. When applying external DC bias, the electrostatic force between the bias pad and the actuator will pull the actuator down and provide an RF signal path. (b) A photomicrograph of a fabricated switch.

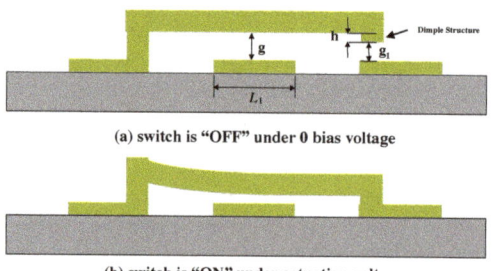

Figure 2. The OFF and ON states of the RF MEMS switch under different DC bias conditions. The cantilever to bias pad gap g and dimple to CPW distance g_1 are both marked. (a) The Switch is under OFF state. (b) The Switch is under ON state.

The first step of this switch design is to choose the desired substrate. Among several widely used materials, high resistivity silicon and fused quartz are chosen. These two substrates have significant differences in physical features that lead to different approaches in their mechanical and electrical designs. Silicon, with its higher relative permittivity ($\epsilon_r = 11.9$) than fused quartz ($\epsilon_r = 3.8$), takes less physical length to build the same RF circuit on a silicon substrate, moreover, the transmission line features are also narrower. This can be an advantage to develop a more compact device but requires finer adjustment on the device RF optimization. On the other hand, depending on the type of fused quartz been used, it could have a dielectric strength as high as 10 MV/cm [14] in comparison with 0.3 MV/cm for resistivity silicon substrates. Therefore, using equivalent circuit dimensions, fused quartz substrate can potentially support 30× higher DC bias voltage than the silicon substrate. Such an additional safe margin provides more design flexibility for designing electrostatic actuated MEMS devices. Meanwhile, high resistivity silicon's resistivity is at the order of 10^4 $\Omega \cdot$cm [15]; in comparison, fused quartz has a higher resistivity at the order of 10^{14}–10^{16} $\Omega \cdot$cm [16]. A thin layer of silicon dioxide should be deposited on top of the silicon surface to improve DC isolation.

The actuator of the proposed RF MEMS switch can be modeled as a cantilever from a mechanical perspective. The cantilever's spring constant k can be calculated by the following equation [17]:

$$k = 2Ew\left(\frac{t}{l}\right)^3 \frac{1 - \left(\frac{x}{l}\right)}{3 - 4\left(\frac{t}{l}\right)^3 + 4\left(\frac{t}{l}\right)^4}. \qquad (1)$$

Its pull-down voltage V_{pull} can be estimated by the following equation [1]:

$$V_{pull} = \sqrt{\frac{8kg^3}{27\epsilon_0 L_1 w}}. \qquad (2)$$

Here, E is Young's modulus of the cantilever material; x is the distance from the cantilever anchor to the center of the bias pad; l, t, and w are the cantilever's length, thickness, and width. As shown in Figures 1 and 2, g and L_1 are the actuation gap and the length of the actuation pad beneath the actuator.

A small dimple structure is attached beneath the cantilever tip. Its thickness h is 0.2 μm. This dimple reduces the contact surface at the switch's ON state to reduce the Van der Waals forces.

Previous research [17] included a similar sized dimple beneath the cantilever beam tip, and the experimental result suggests a restoring force of 0.07 mN is needed for reliable restoration. Assuming gold as the cantilever material, l in a range of 70–100 μm and x is roughly half of l, w and t can be calculated accordingly. In consideration of fabrication feasibility, w should be at least 2–3 times longer than t. With an actuation gap g of 1.2 μm, the initial values of t and w are selected as 2.3 μm and 7 μm, respectively. The resulting V_{pull} is estimated to be around 55 V.

To prevent dielectric breakdown during actuation, the electric field between the bias pad and nearby CPW should be kept well below the substrate dielectric strength. As presented in Figure 3, a crude estimate of the electric field strength can be obtained by assuming a uniform field distribution.

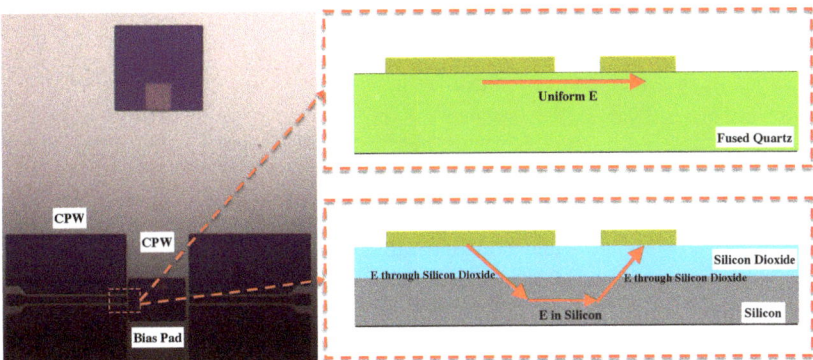

Figure 3. The strongest electric field is between the CPW and the adjacent DC bias pad.

In the fused quartz-based design, the strongest electric field extends along the minimum gap between the CPW and adjacent bias pad. Under a 55-V bias voltage, the electric field is approximately 0.11 MV/cm along the 5 µm gap. This value is well below the substrate's dielectric strength and can be further reduced by increasing the gap between the bias pad and CPW. By comparison, in the silicon based design, the electric field extended through the silicon dioxide thickness as depicted in Figure 3. Considering silicon dioxide's dielectric strength of around 3–5 MV/cm, the silicon dioxide insulation layer should be around 100 nm to prevent dielectric breakdown. One common method to deposit a silicon dioxide layer is through plasma-enhanced chemical vapor deposition (PECVD). However, previous fabrication and subsequent tests suggest this PECVD oxide has a high risk of creating an un-covered silicon area known as pinholes inside the oxidation layer that result in potential shorts between the MEMS devices and silicon substrate [18,19]. In this work, a very uniform dry thermal silicon-dioxide layer was grown on the high resistivity silicon surface with a thickness of 100–110 nm.

3. Electrical Design

In Figure 4, a cross-section view of a coplanar waveguide (CPW) structure is presented. The impedance of a CPW is largely determined by the signal line's width, w, and the signal to ground gap, g_c. When scaling w and g_c by the same ratio, CPW's impedance has little variation. At a minimum feature size $g_c = 4$ µm, the corresponding w to realize a 50 Ω impedance on different substrates is simulated through Ansys HFSS and listed in Table 1. The result shows that a lower relative permittivity substrate require a large signal line width. On silicon, the required signal line width is 7 µm, while for a fused quartz, the signal line width increases to 35 µm.

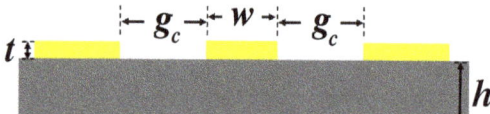

Figure 4. The cross-section view of a CPW. In this work, both high resistivity silicon and fused quartz substrate have thickness h of 500 µm. The CPW is evaporated gold with a thickness t of 400 nm. The signal line width w and signal-ground gap g are impacted by the substrate's relative permittivity.

Table 1. The cross-section dimensions of a 50 Ω CPW on different substrates.

Substrate Material	Relative Permittivity ϵ	g (µm)	w (µm)
Silicon	11.9	4	7
SiC	9.7	4	9
AlN	9.2	4	10
Quartz	4.0	4	35

The dimensions of 50 Ω CPWs on the different substrates are compared through Ansys HFSS simulation. In the CPW design, the signal-ground gap g_c is limited to 4 µm to keep the transmission line fabrication feasible. With the same g_c, the differences in signal line width w are compared under different relative permittivity ϵ.

At the OFF state, the switch's actuator is coupled to the bias pad and the CPW signal line through parasitic capacitances as presented in Figure 5. Here C_1 is actuator-bias pad coupling capacitance; C_2 is pad-CPW coupling capacitance; C_{tip} is the coupling capacitance between actuator tip and CPW. The equivalent series capacitance of C_1 and C_2 is designed to be much smaller than C_{tip} so that the total parasitic capacitance $C_{total} \approx C_{tip}$. At the OFF state, the switch's RF isolation is determined by the impedance of the total capacitance.

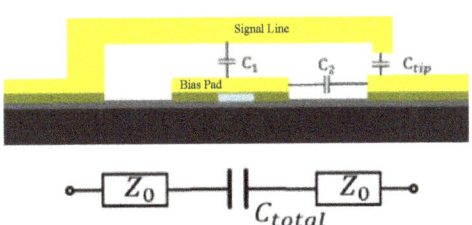

Figure 5. The parasitic capacitance of MEMS switch under OFF state and the switch's equivalent circuit model.

To reach a higher RF isolation, a smaller C_{tip} is desired. Assuming the switch actuator and CPW overlaps by 2 μm, the initial switch designs on silicon and fused quartz are each shown in Figures 6a and 7a. Using the parallel plate capacitance equation, the tip capacitance of the silicon and quartz designs are estimated to be 0.6 fF and 2 fF, respectively. From these initial capacitance values, their isolation performance is simulated through a simplified transmission line circuit model. The results are plotted in blue and black curves in Figure 8. In this work, limited by equipment availability, the highest measurement frequency is 750 GHz; so the simulation is plotted in the DC-750 GHz frequency range. For the purposes of switch design comparison, a 15 dB isolation level is drawn as a reference. The estimation suggests the silicon design will have 15 dB isolation at around 470 GHz; due to the wider cantilever causing larger capacitance, the isolation of the quartz design is reduced to 15 dB at about 140 GHz.

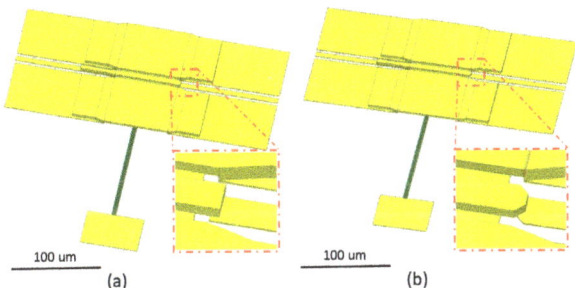

Figure 6. Three dimensional (3-D) model of high resistivity silicon-based MEMS switch. The initial design (**a**) has a larger parallel area at the actuator tip, which causes larger capacitance. To reduce such capacitance for higher isolation performance, its actuator tip is trimmed (**b**).

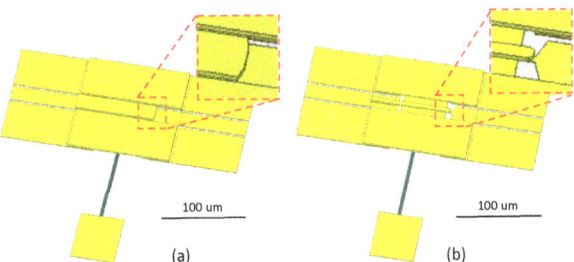

Figure 7. Three dimensional (3-D) model of fused quartz-based MEMS switch. In the initial design (**a**) the actuator has the same width as the CPW's signal line, which causes roughly 2 fF parasitic capacitance at the tip. To reduce such capacitance, the actuator's width was reduced and the tip is trimmed to further reduce the capacitance (**b**).

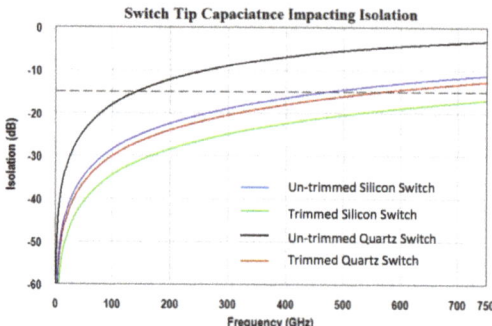

Figure 8. Estimated switch isolation performance dominated by actuator tip capacitance.

To improve the isolation performance, the overlapping area that forms C_{tip} must be trimmed to minimize the cantilever's parasitic capacitance. In the silicon switch, as shown in Figure 6b, the cantilever tip and the CPW's signal line tip are both tapered from 7 µm to 3.5 µm. With a smaller overlapping area, the capacitance is reduced to 0.3 fF. In the quartz switch, the cantilever width is reduced to 8 µm. To match the elevated CPW impedance to 50 Ω, the horizontal gap between signal and ground is adjusted to 15 µm; the CPW's signal line tip is also tapered to 3 µm. This adjustment is presented in Figure 7b. Through trimming the geometries of the cantilever and the CPW, the parasitic capacitance of the quartz switch is significantly reduced to roughly 0.5 fF. The estimated RF isolation of both silicon and fused quartz switches are plotted in green and red in Figure 8. The fused quartz switch's cut-off frequency with 15 dB isolation is expanded to 560 GHz, in comparison, the silicon switch's 15 dB isolation bandwidth is higher than 750 GHz.

At the ON state, the switch's RF performance can also be analyzed through a transmission line model. As presented in Figures 1 and 2, the actuator in the MEMS switch creates a discontinuity in physical geometry and impedance mismatch. The switch's actuator and its adjacent ground lines can be modeled as a CPW section, which is elevated from the substrate. As the CPW elevates further away from the silicon substrate surface, the effective relative permittivity ε_{eff} for the CPW decreases and the impedance increases. This relationship is presented in Figure 9a. In Table 2, the simulated CPW impedance with different elevation heights above the silicon substrate is provided.

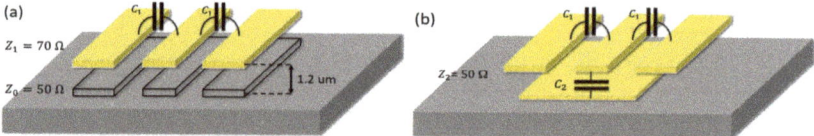

Figure 9. (a) Impedance variation caused by CPW structure elevation. (b) impedance tuning is realized by adjusting the DC bias pad size.

Table 2. CPW impedance & its elevation height above the silicon substrate.

Elevation Height (µm)	Impedance (Ω)
0	50
0.4	63.6
0.8	64.3
1.2	70

Because the actuator beam is elevated to 1.2 µm above the substrate, the effective relative permittivity ε_{eff} reduces to around 4.8, the impedance rises to 70 Ω, and the return loss will be more than 10 dB at 700 GHz. Such RF performance change can be explained through a simplified transmission line model in Figure 10. For example, at 3 GHz, a 70 μm cantilever beam has an electrical length of merely 0.9° and the resulting impedance mismatch is negligible. In comparison, at 300 GHz, the electrical length significantly increases to 90°, which naturally has a significant impact on the RF performance.

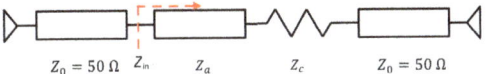

Figure 10. Simplified transmission line model of MEMS switch at the ON state.

The cantilever's impedance mismatch is most significant at the frequency when it reaches 90° electric length. Here Z_c is the introduced resistance due to switch Ohmic contact, on a scale of around 1 Ω or less. The Z_a section is a quarter wavelength long and therefore, the input impedance at is given by:

$$Z_{in} = Z_a^2 / Z_0$$

In this case, the equivalent impedance of Z_{in} is 98 Ω.

In order to reduce this impedance mismatch, the DC bias pad size is engineered to reduce the actuator impedance Z_a. Since the coplanar waveguide's characteristic impedance is approximated as:

$$Z = \sqrt{\frac{L}{C'}}, \tag{3}$$

by increasing the bias pad's size, the total parasitic capacitance at the actuator area is doubled, and the impedance is then successfully tuned to 48 Ω. This engineering tuning method is presented in Figure 9b.

A more comprehensive circuit model includes the anchor and the actuator's tip is provided in Figures 11 and 12 respectively. Figure 11 represents the ON state, in which the actuator is pulled down to make tip contact at the free end. Z_{anc} represents the anchor section of the actuator; Z_a represents the elevated actuator after impedance tuning. Z_{elev} represents the elevated CPW section assuming no fringing capacitance impact from the bias pad. The switch's transmission line model for the OFF state is shown in Figure 12. The gap beneath the actuator tip serves as an isolation capacitor and introduces impedance Z_{c1}.

The simulation result of a silicon-based switch at the ON state is presented in Figure 13a. In this figure, both the transmission line model and HFSS finite element analysis are included and compared. The insertion loss is smaller than 1 dB across DC-750 GHz band, while the return loss is better than 15 dB. Compared with EM simulation, the transmission line model catches the major scale and trends of insertion loss and return loss.

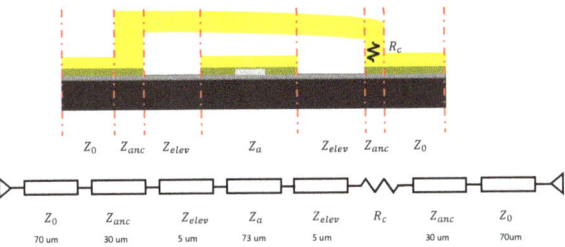

Figure 11. A comprehensive transmission line circuit model for ON state MEMS switch.

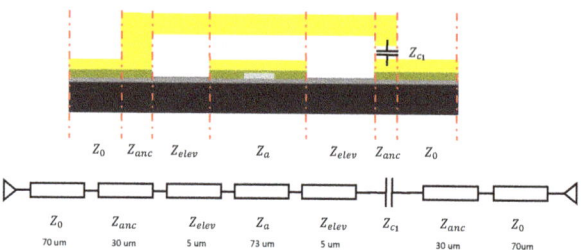

Figure 12. A comprehensive transmission line circuit model for OFF state MEMS switch.

The silicon switch's OFF state simulation is given in Figure 13b. The result shows the expected OFF state isolation better than 15 dB. The transmission line model fits the EM simulation result well in the isolation plots (S21/S12). In the reflection plots (S11/S22), the transmission line model does not account for the radiated energy. As a result, its simulation result has a minor difference from HFSS analysis. Similar results are also seen in fused quartz based simulation in Figure 13c,d.

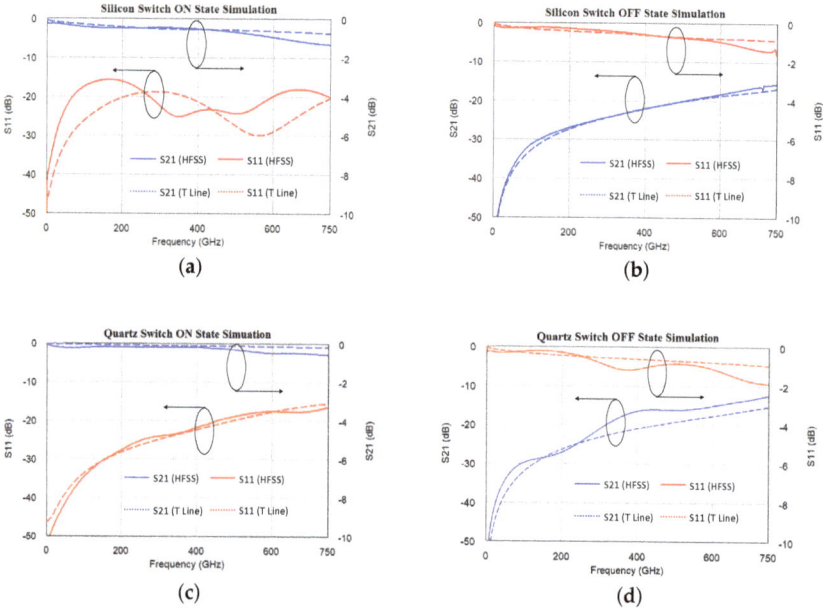

Figure 13. The THz MEMS switches were modeled as transmission line circuits and simulated using AWR Microwave Office (marked as T line). The circuit simulation results are compared with ANSYS HFSS finite element analysis. (**a**) Silicon based design at the ON state. (**b**) Silicon based design at the OFF state. (**c**) Quartz based design at the ON state. (**d**) Quartz based design at the OFF state.

4. Switch Fabrication Challenge & Solution

The silicon-based MEMS switch fabrication flow is simplified and presented in Figure 14. The fabrication process starts with circuit layer deposition using the lift-off technique. After that, two aluminum sacrificial layers are deposited and the dimple position is prepared with another lift-off. Dry etching and plating were used to form the anchor of switch. After the gold seed layer and photoresist patterning, the actuator beam was plated. After a series of wet etch steps, the switch is finally released by a critical point dryer (CPD). The quartz-based MEMS switch fabrication process is very similar.

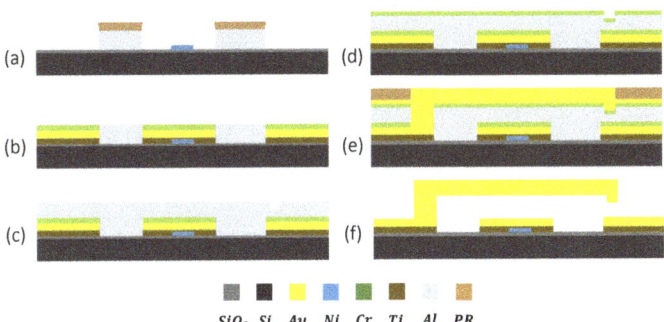

Figure 14. A simplified silicon-based MEMS switch fabrication flow. The fabrication process starts with circuit layer deposition using a lift-off technique in (**a**,**b**). After that, two aluminum sacrificial layers are deposited and dimple position is prepared with another lift-off in (**c**). Etching and plating were used to form the switch's anchor in (**d**). After the gold seed layer and a photoresist patterning, the beam was plated in (**e**). After a series of wet etch steps, the switch is finally released by CPD in (**f**).

Typical MEMS fabrication uses photoresists such as SU-8 as a sacrificial layer, which has a coefficient of thermal expansion (CTE) as high as 50–102 ppm/K [20,21]. In comparison, Aluminum's CTE is 25.5 ppm/K, much closer to Gold's CTE of 13.9 ppm/K. In this research, using an Al sacrificial layer significantly reduced the CTE mismatch between the actuator beam and the sacrificial layer beneath it, which prevented the commonly observed beam bowing after device release [22].

The Al sacrificial layer can also be easily removed through halogen gas-based reactive ion etch (RIE) or alkali solution wet etch. However, using an Al sacrificial layer also brings challenges in fabrication because it can easily form Al-Au compounds which cannot be easily removed. This red Al-Au compound can be observed in Figure 15a. A chromium layer is added as a diffusion barrier layer between the Al sacrificial layer and Au circuit layer [23,24]. In Figure 14d–e, a chlorine RIE and wet etch combined procedure is used to remove the aluminum layer and the chromium barrier layer.

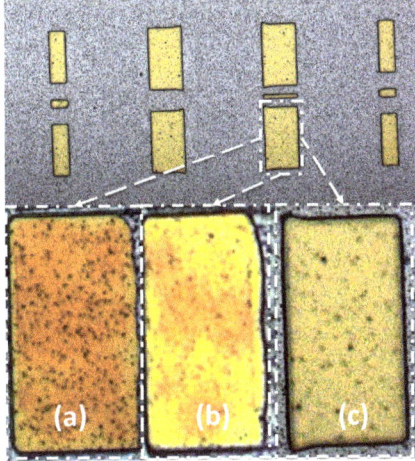

Figure 15. (**a**) The Al-Au compounds observed (**b**) Al-Au compounds are reduced with 50 nm chromium barrier layer applied between the gold and aluminum layer (**c**) 80 nm chromium barrier layer prevented Al-Au to produce.

5. Switch Calibration & Measurement

In order to verify the switch design, the RF MEMS switch is fabricated and tested. An on-wafer measurement is set up as presented in Figure 16. A Keysight VNA (PNA-5245A) is used to conduct a two-port S-parameter measurement. Four sets of VNA extender pairs (VDI WR-5.1, WR3.4, WR2.2, and WR-1.5) are used to up-convert VNA test frequency sweeping to 140–220 GHz, 220–330 GHz, 330–500 GHz, and 500–750 GHz. To conduct the on-wafer measurement, DMPI T-wave probes (WR5.1, WR3.4, WR2.2, and WR1.5) are also used correspondingly [25]. An SEM image of the calibration kit and completed RF MEMS contact switch is provided in Figure 17. The switch's DC high voltage bias and ground are provided through a bias tee integrated into the DMPI probes. A series 10 MΩ resistance is placed in the DC bias circuit to limit the current below 10 µA in the potential breakdown condition.

Before measurement, on-wafer calibration is applied using Through-Reflection-Line (TRL) standards and WinCal XE calibration software. Standards include a 120 µm through, two reflection and three lines. The three different lines have additional lengths of 45 µm, 67 µm and 113 µm. Redundant reflections and lines are used for higher calibration accuracy [25]. Calibrated measurement shows the loss of 50 Ω CPW is around 3.5 dB/mm at 500 GHz and around 6.5 dB/mm at 750 GHz.

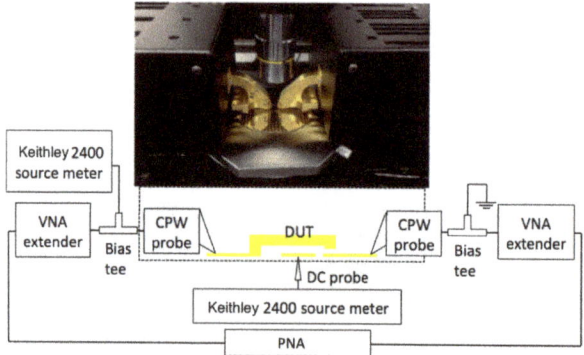

Figure 16. On wafter two-ports probing set up used for MEMS switch RF measurement is shown in this diagram.

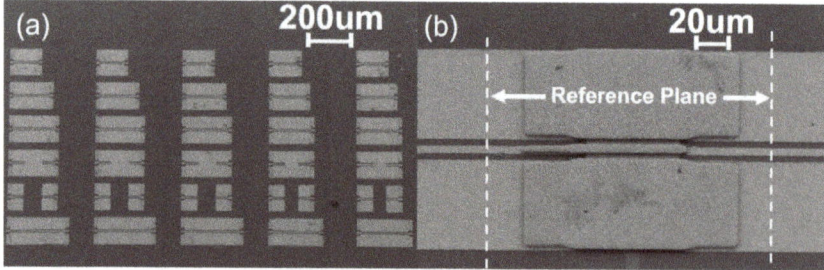

Figure 17. SEM image of (**a**) on-wafer TRL calibration kit and (**b**) an example silicon switch.

The silicon switch measurement is compared with corresponding HFSS simulation results in Figure 18a,b. The ON state performance is shown in Figure 18a. The measurement shows the switch's impedance is well matched and has a small variation over frequency. The return loss is better than 10 dB in the whole band. The switch's insertion loss is thus dominated by the actuator contact resistance and metal loss. Previous discussion demonstrated the contact resistance will bring insertion loss merely on a scale of 0.1 dB across the whole band; meanwhile, the metal loss will cause switch insertion loss to increase slightly over frequency. The measured insertion loss (S21 and S12) in Figure 18a fits the expectation. The return loss (S11 and S22) is mostly better than 20 dB as HFSS simulation suggested. Because the measurement is conducted in four different waveguide bands through four calibrations, there are inevitably certain minor measurement errors at the boundary frequency points. However, the return loss measurement trend still fits the simulated curve.

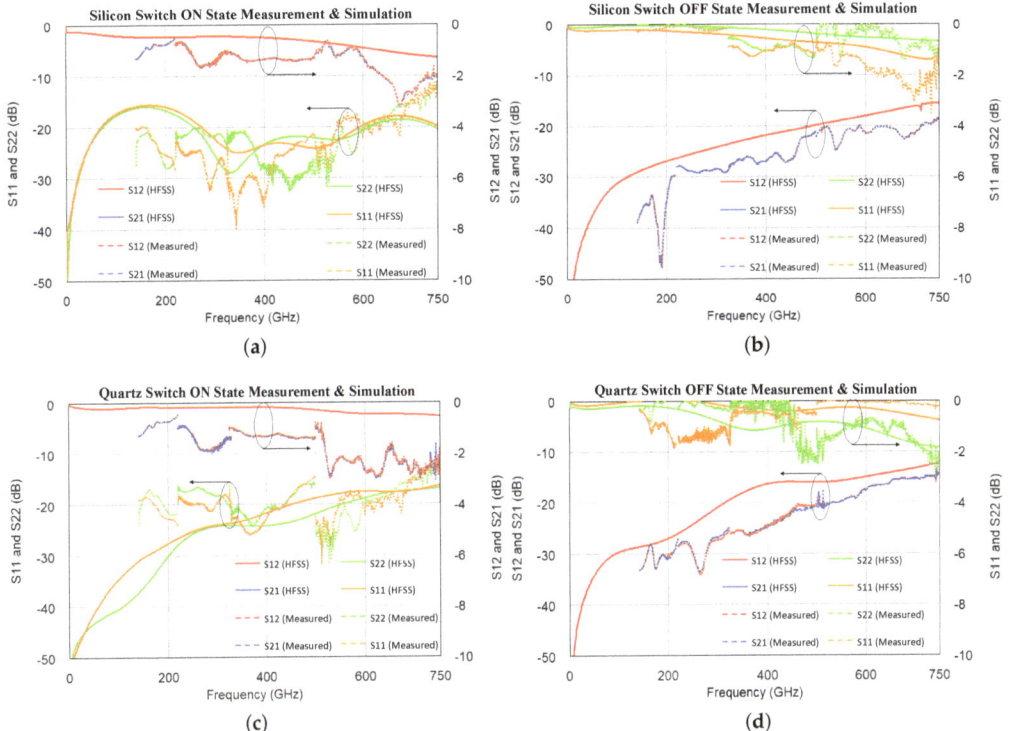

Figure 18. The THz MEMS switch HFSS simulation measurement comparison. (**a**) Silicon switch ON state measurement and simulation. (**b**) Silicon switch OFF state measurement and simulation. (**c**) Quartz switch ON state measurement and simulation. (**d**) Quartz switch OFF state measurement and simulation.

The silicon switch's OFF state measurement is presented in Figure 18b. Because the CPW sections are isolated by the air-gap capacitance, any incident RF energy is mostly reflected. In this measurement, the return loss (S11 and S22) curves are only slightly lower than 0 dB, which fits the simulated result well. The isolation plot (S12 and S21) also match the simulated curves.

A similar condition is also observed in the fused quartz switch's ON and OFF state measurements. Comparing Figure 13a,c, the silicon switch's impedance match is expected to be better than the fused quartz switch in the 200–500 GHz band, which is due to different bias pad-actuator coupling conditions and associated resonance. The measurement in Figure 18a,c matches such expectation. Meanwhile, the fused quartz switch and silicon switch have similar insertion loss conditions, which is very reasonable. In both switch designs, insertion loss is majorly determined by the actuator's contact resistance and gold metal loss. In both switches, the actuation bias voltage is 55 V, and both actuators have the same thickness. The actuation force is expected to be similar, which leads to the same expected contact resistance. Gold is used in both switch designs, which provides the same metal loss over frequency. Those factors determine both switches to have fairly similar and constant insertion loss over frequency; as frequency increases, the insertion loss increases by a similar magnitude in both silicon and fused quartz designs.

The quartz switch measurement in Figure 18d also follows the HFSS simulation result well. Similar to the silicon switch's OFF state measurement, the fused quartz switch has 0.3–0.4 fF air gap capacitance. The impedance of this capacitance is dominant compared with 50 Ω transmission line impedance. As frequency increases, the reduced capacitance impedance leads to reduced isolation and rising S21/S12 measurement curves over frequency.

6. Discussion & Conclusions

In this paper, a THz MEMS switch design process is presented. To validate this process, THz MEMS switches are realized on both silicon and fused quartz as examples of both high and low dielectric constant substrates respectively.

In the device design, electrostatic actuation is selected to control the switches in consideration of its advantage over device size, integration challenges, switching speed, power consumption, and RF performance. Manufacturing limits and mechanical reliability together determined the minimum dimensions and thus the switches' geometries. To integrate elevated actuators with CPW, the bias pad geometry is engineered to provide an optimized impedance match. The designed MEMS switches are modeled through transmission line analysis as well as finite element-based electromagnetic simulations. The comparison suggests the transmission line model captures the major electrical features successfully; meanwhile, the finite element model can also evaluate certain minor coupling, resonance, and frequency-dependent conductor loss. Fabrication flow of the THz MEMS switch is provided, in which a diffusion barrier layer is used to prevent forming Au-Al compound [26]. A new RIE/wet-etch combined process is critical to selectively etch certain metal layers. Both silicon and fused quartz switches are calibrated through the two-port TRL method from 140 to 750 GHz. The measurements fit previous modeling and simulation results well and serve to verify this THz MEMS switch design process.

Author Contributions: Conceptualization, Y.F.; methodology, Y.F.; software, Y.F.; validation, Y.F. and H.-y.T.; formal analysis, Y.F.; investigation, Y.F. and N.S.B.; data curation, Y.F.; writing-original draft, Y.F.; writing-review and editing, N.S.B.; supervision, N.S.B.; project administration, N.S.B. All authors have read and agreed to the published version of the manuscript.

Funding: The authors wish to acknowledge the U.S. National Ground Intelligence Center (NGIC) for their financial support of this work through contract W911W5-11-C-0013.

Conflicts of Interest: The authors declare no conflict of interest.

Abbreviations

The following abbreviations are used in this manuscript:

MEMS	Micro-electromechanical systems
CPW	Coplanar waveguide
CMOS	Complementary metal-oxide-semiconductor
BiCMOS	Bipolar Complementary metal-oxide-semiconductor

References

1. Rebeiz, G.M. *RF MEMS: Theory, Design, and Technology*; John Wiley & Sons: Hoboken, NJ, USA, 2004.
2. Brown, E. RF-MEMS switches for reconfigurable integrated circuits. *IEEE Trans. Microw. Theory Tech.* **1998**, *46*, 1868–1880. [CrossRef]
3. Entesari, K.; Rebeiz, G. A differential 4-bit 6.5-10-GHz RF MEMS tunable filter. *IEEE Trans. Microw. Theory Tech.* **2005**, *53*, 1103–1110. [CrossRef]
4. Sekar, V.; Armendariz, M.; Entesari, K. A 1.2–1.6-GHz Substrate-Integrated-Waveguide RF MEMS Tunable Filter. *IEEE Trans. Microw. Theory Tech.* **2011**, *59*, 866–876. [CrossRef]
5. Cho, Y.H.; Rebeiz, G.M. Two- and Four-Pole Tunable 0.7–1.1-GHz Bandpass-to-Bandstop Filters With Bandwidth Control. *IEEE Trans. Microw. Theory Tech.* **2014**, *62*, 457–463. [CrossRef]

6. Schoenlinner, B.; Abbaspour-Tamijani, A.; Kempel, L.; Rebeiz, G. Switchable low-loss RF MEMS Ka-band frequency-selective surface. *IEEE Trans. Microw. Theory Tech.* **2004**, *52*, 2474–2481. [CrossRef]
7. Keimel, C.; Claydon, G.; Li, B.; Park, J.N.; Valdes, M.E. Microelectromechanical-Systems-Based Switches for Power Applications. *IEEE Trans. Ind. Appl.* **2012**, *48*, 1163–1169. [CrossRef]
8. Goggin, R.; Fitzgerald, P.; Stenson, B.; Carty, E.; McDaid, P. Commercialization of a reliable RF MEMS switch with integrated driver circuitry in a miniature QFN package for RF instrumentation applications. In Proceedings of the 2015 IEEE MTT-S International Microwave Symposium, Phoenix, AZ, USA, 17–22 May 2015; pp. 1–4. [CrossRef]
9. Sim, S.M.; Lee, Y.; Jang, Y.H.; Lee, Y.S.; Kim, Y.K.; Llamas-Garro, I.; Kim, J.M. A 50–100 GHz ohmic contact SPDT RF MEMS silicon switch with dual axis movement. *Microelectron. Eng.* **2016**, *162*, 69–74. [CrossRef]
10. Du, Y.; Su, W.; Tolunay, S.; Zhang, L.; Kaynak, M.; Scholz, R.; Xiong, Y.Z. 220 GHz wide-band MEMS switch in standard BiCMOS technology. In Proceedings of the 2015 IEEE Asian Solid-State Circuits Conference (A-SSCC), Xiamen, China, 9–11 November 2015; pp. 1–4.
11. Wipf, S.T.; Göritz, A.; Wipf, C.; Wietstruck, M.; Burak, A.; Türkmen, E.; Gürbüz, Y.; Kaynak, M. 240 GHz RF-MEMS switch in a 0.13 µm SiGe BiCMOS Technology. In Proceedings of the 2017 IEEE Bipolar/BiCMOS Circuits and Technology Meeting (BCTM), Miami, FL, USA, 19–21 October 2017; pp. 54–57.
12. Shah, U.; Reck, T.; Decrossas, E.; Jung-Kubiak, C.; Frid, H.; Chattopadhyay, G.; Mehdi, I.; Oberhammer, J. 500–750 GHz submillimeter-wave MEMS waveguide switch. In Proceedings of the 2016 IEEE MTT-S International Microwave Symposium (IMS), San Francisco, CA, USA, 22–27 May 2016; pp. 1–4.
13. Feng, Y.; Barker, N.S. High performance 500–750 GHz RF MEMS switch. In Proceedings of the 2017 IEEE MTT-S International Microwave Symposium (IMS), Honolulu, HI, USA, 4–9 June 2017; pp. 1095–1097.
14. Kasap, S.O.; Capper, P. *Springer Handbook of Electronic and Photonic Materials*; Springer: Berlin, Germany, 2006; Volume 11.
15. Reyes, A.C.; El-Ghazaly, S.M.; Dorn, S.; Dydyk, M.; Schroder, D.K.; Patterson, H. High resistivity Si as a microwave substrate. In Proceedings of the 1996 46th Electronic Components and Technology Conference, Orlando, FL, USA, 28–31 May 1996; pp. 382–391.
16. Nayak, P. Characterization of High-Resistivity Silicon Bulk and Silicon-On-Insulator Wafers. Ph.D. Thesis, Arizona State University, Phoenix, AZ, USA, 2012.
17. Gong, S. DC-Contact Cantilever RF-MEMS Switches at Millimeter-Wave Frequencies. Ph.D. Thesis, University of Virginia, Charlottesville, VA, USA, 2010.
18. Liu, H.; Song, X.; Wang, Q.; Wu, W.; Fan, J.; Liu, J.; Tu, L. Healing pinhole shorts for applications using intermetal dielectric films. In Proceedings of the 2017 IEEE 12th Nanotechnology Materials and Devices Conference (NMDC), Singapore, 2–4 October 2017; pp. 68–69.
19. Woodson, T. Characterization of the Silicon Dioxide Film Growth by Plasma Enhanced Chemical Vapor Deposition (PECVD). Princeton Plasma Physics Laboratory Summer Program 2004. Available online: https://www.google.com.sg/url?sa=t&rct=j&q=&esrc=s&source=web&cd=&ved=2ahUKEwja7ICE_c73AhUhqFYBHdmMB_AQFnoECAMQAQ&url=https%3A%2F%2Fw3.pppl.gov%2Fppst%2Fdocs%2Fwoodson.pdf&usg=AOvVaw12HG5_xUhg-y7Ym7Z3vSUl (accessed on 28 March 2022).
20. Lorenz, H.; Despont, M.; Fahrni, N.; LaBianca, N.; Renaud, P.; Vettiger, P. SU-8: A low-cost negative resist for MEMS. *J. Micromech. Microeng.* **1997**, *7*, 121. [CrossRef]
21. Feng, R.; Farris, R.J. The characterization of thermal and elastic constants for an epoxy photoresist SU8 coating. *J. Mater. Sci.* **2002**, *37*, 4793–4799. [CrossRef]
22. Stanec, J.R.; Begley, M.R.; Barker, N.S. Mechanical properties of sacrificial polymers used in RF-MEMS applications. *J. Micromech. Microeng.* **2006**, *16*, 2086. [CrossRef]
23. Shen, H.; Gong, S.; Barker, N.S. DC-Contact RF MEMS Switches using Thin-Film Cantilevers. In Proceedings of the 2008 European Microwave Integrated Circuit Conference, Amsterdam, The Netherlands, 27–28 October 2008; pp. 382–385. [CrossRef]
24. Gong, S.; Shen, H.; Barker, N.S. A 60-GHz 2-bit Switched-Line Phase Shifter Using SP4T RF-MEMS Switches. *IEEE Trans. Microw. Theory Tech.* **2011**, *59*, 894–900. [CrossRef]
25. Reck, T.J.; Chen, L.; Zhang, C.; Arsenovic, A.; Lichtenberger, A.; Weikle, R.M.; Barker, N.S. Calibration accuracy of a 625 GHz on-wafer probe. In Proceedings of the 2010 76th ARFTG Microwave Measurement Conference, Clearwater Beach, FL, USA, 30 November–3 December 2010; pp. 1–5.
26. Gong, S.; Shen, H.; Barker, N.S. Study of Broadband Cryogenic DC-Contact RF MEMS Switches. *IEEE Trans. Microw. Theory Tech.* **2009**, *57*, 3442–3449. [CrossRef]

Room-Temperature CMOS Monolithic Resonant Triple-Band Terahertz Thermal Detector

Xu Wang [1,2,3], Ting-Peng Li [1], Shu-Xia Yan [2,3,*] and Jian Wang [1,3,4,*]

1. State Key Laboratory of Complex Electromagnetic Environmental Effects on Electronics and Information System, Luoyang 471003, China
2. School of Electronics and Information Engineering, Tiangong University, Tianjin 300387, China
3. School of Microelectronics, Tianjin University, Tianjin 300072, China
4. Qingdao Institute for Ocean Technology, Tianjin University, Qingdao 266200, China
* Correspondence: yanshuxia@tiangong.edu.cn (S.-X.Y.); wangjian16@tju.edu.cn (J.W.)

Abstract: Multiband terahertz (THz) detectors show great application potential in imaging, spectroscopy, and sensing fields. Thermal detectors have become a promising choice because they could sense THz radiations on the whole spectrum. This paper demonstrates the operation principle, module designs with in-depth theoretical analysis, and experimental validation of a room-temperature CMOS monolithic resonant triple-band THz thermal detector. The detector, which consists of a compact triple-band octagonal ring antenna and a sensitive proportional to absolute temperature (PTAT) sensor, has virtues of room-temperature operation, low cost, easy integration, and mass production. Good experimental results are obtained at 0.91 THz, 2.58 THz, and 4.2 THz with maximum responsivities of 32.6 V/W, 43.2 V/W, and 40 V/W, respectively, as well as NEPs of 1.28 $\mu W/Hz^{0.5}$, 2.19 $\mu W/Hz^{0.5}$, and 2.37 $\mu W/Hz^{0.5}$, respectively, providing great potential for multiband THz sensing and imaging systems.

Keywords: terahertz; multiband detector; CMOS; octagonal ring antenna

1. Introduction

Terahertz (THz) waves possess many unique properties, such as penetrating non-conductive materials, which are opaque in visible and infrared bands [1]. They could identify specific materials according to their characteristic THz signatures [2]. THz waves are safe for biological tissue because of low photon energy and non-ionizing attributes, in contrast to X-rays, and promise higher resolution compared with microwave bands [3,4]. The above characteristics of THz waves have promoted THz technology to make great progress in medical detection [5], security inspection [6], non-destructive testing [7], wireless communication [8], atmospheric monitoring [9], astronomical observation [10], and so on. However, the interest in a wide range of commercial THz applications is the main driver for the development of widely accessible room-temperature THz detectors. In addition, as the same material is imaged at different frequencies, the image sharpness would vary with different transmission rates. Hence, multiband detectors could dramatically improve the overall sensing and imaging ability by means of obtaining more informative images through fusion technology [11]. Besides, multiband detectors also have the advantages of enhanced detection probability, increased calibration capability, and reduced influences of standing waves or scattering, showing great potential for further development of THz applications [11–14].

With continuous developments in CMOS technology, which is considered as an attractive device technology because of its low cost, high yield, and high integration ability [15], various kinds of room-temperature CMOS multiband THz detectors have attracted more attention and have gradually received in-depth research in the last couple of years [16–20].

Multiband active detectors use a higher harmonic, resulting in sharply increased noise figures and fixed operation frequencies, which are determined by fundamental and harmonic frequencies [21]. Similar trends do not exist for passive devices that are more suitable for human vision and image processing [22]. Therefore, several multiband passive detectors consisting of antennas and MOSFETs have been proposed [16–20]. The above FET-based THz detectors, whose operation frequencies are seriously restricted and influenced by transistors, achieve better characteristic results below 1 THz, and their performances degrade dramatically as the frequency exceeds 1 THz owing to frequency-dependent parasitic elements [19]. Besides, the detectors in [19,20] are composed of multiple discrete antennas and multiple FETs, resulting in lower integration levels, a larger chip area, and higher cost, thus it is necessary to design compact multiband THz detectors. However, it is hard to obtain compact FET-based detectors because the input impedance of transistors differs at multiple frequencies [16,21]. Compared with FET-based detectors, THz thermal detectors constitute promising options as they allow wideband detection, support high-frequency THz detection, and show performance advantages in higher THz bands because their output signals are independent of frequencies [23–25]. Therefore, several room-temperature CMOS multiband THz thermal detectors have been proposed. A room-temperature CMOS multiband THz thermal detector composed of an antenna and an NMOS sensor is proposed, but it operates at 0.546 THz, 0.688 THz, 0.78 THz, and 0.912 THz [26]. It is necessary to design room-temperature CMOS multiband THz thermal detectors that could detect sub 1 THz waves and above 1 THz waves to possess good sensitivity and high resolution [11]. Previous works have described two kinds of CMOS triple-band THz thermal detectors, which mainly concentrate on modules' designs, including designs of receiving structures and temperature sensors, thus they lacked the concept of collaborative designs between modules, such as completing the layout of a temperature sensor according to the raised temperature distribution of receiving structures [27,28]. Besides, these triple-band detectors just completed the performance characterizations at two frequencies.

This paper presents a compact room-temperature triple-band THz thermal detector made up of a strong octagonal ring antenna and a sensitive PTAT sensor using a Global Foundry 55 nm CMOS process. Because lower THz waves have a greater penetration depth and higher THz waves provide better spatial resolution, the proposed detector is chosen to operate at 0.91 THz, 2.58 THz, and 4.2 THz for available THz sources so as to obtain better penetration and greater spatial resolution. It achieves relatively better measurement results at three operation frequencies with detailed analysis, exactly presenting an uncooled, compact, cost-effective, easy-integration, and mass-production multiband detection system.

2. Detector Structure and Operation Principle

Antennas are ready to shape the radiation pattern and tune the impedance match within a wider bandwidth [29], while octagonal rings are used to constitute antennas because they have advantages of smaller chip area occupation and less coupling effect than other structures [30]. In addition, PTAT sensors as a type of common CMOS temperature sensor show great application potential owing to their better linearity and accuracy [31]. Based on this, Figure 1 shows the schematic diagram of the proposed detector, which consists of a compact triple-band octagonal ring antenna, a polysilicon resistor at the termination of the antenna, and a sensitive PTAT sensor. As THz waves interact with the antenna, an instantaneous frequency-dependent current is excited and flows through the resistor; by this means, incident THz waves are frequency-selective absorbed. Thus, electromagnetic (EM) energy is immediately transformed into thermal energy through ohmic loss and conductive loss, leading to the localized temperature increment depending on the magnitude of the radiation [24]. In addition, the PTAT sensor transforms the rising temperature into an increased output voltage, so the sensor is located below the antenna and in close proximity to the resistor in order to reduce heat loss and sense an increased temperature as fast as possible. The triple-band detection is accomplished as THz waves of three frequencies are incident on the detector successively.

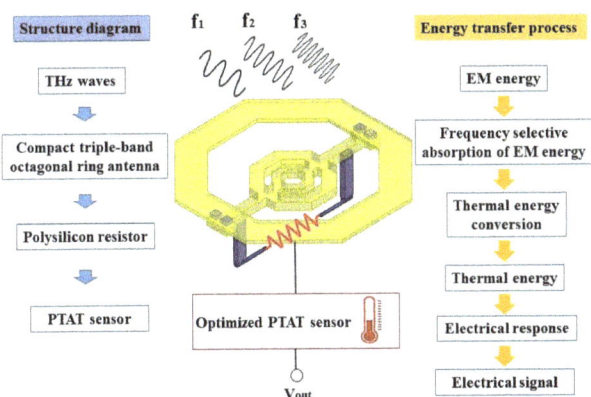

Figure 1. Schematic diagram of the proposed triple-band detector.

Furthermore, a lower operating frequency leads to larger antenna sizes, so the temperature distribution caused by conductor loss is far from that caused by ohmic loss. The resistor becomes the main heat source and presents a strong, uniform, and raised temperature distribution in a certain area, because temperature sensing elements of the sensor should sense the same temperature and the received EM energy is mainly converted into joule heat through the resistor. Therefore, the increased temperature is approximately equal to the temperature increment generated by the resistor. However, a higher operating frequency leads to smaller antenna sizes, so the temperature distribution generated by the antenna and the resistor is closed or even overlapped and, finally, a strong, uniform, and raised temperature distribution is generated in a certain area centered on the resistor. In this way, the perceived temperature increment is approximately equal to the sum temperature increment caused by the resistor and the conductor of the antenna.

According to the operation principle of the detector, its design task not only includes the independent design of the antenna and the PTAT sensor, but also contains the co-design between the antenna and the PTAT sensor based on the temperature distribution of the antenna caused by the incident THz waves.

2.1. Design of Octagonal Ring Antenna

The structure diagram and optimized geometric parameters of a compact triple-band octagonal ring antenna (sample A) using HFSS tools are shown in Figure 2a,b. Nested octagonal rings are composed of three concentric rings with different sizes, and the smaller ring is embedded in the larger ring. As perimeters of the outer ring, the middle ring, and the inner ring are about dielectric wavelengths of 0.91 THz waves, 2.58 THz waves, and 4.2 THz waves, respectively, the fundamental modes of the outer octagonal ring, the middle octagonal ring, and the inner octagonal ring could radiate 0.91 THz waves, 2.58 THz waves, and 4.2 THz waves, correspondingly. Based on the reciprocity theorem, sample A could also receive THz waves of 0.91 THz, 2.58 THz, and 4.2 THz, respectively.

As shown in Figure 2a,b, sample A was made up of three nested octagonal rings, connection structures, a ground plane, two transmission lines, and a grounded wall. The outer octagonal ring was constructed in the metal 9 layer of the 55 nm CMOS process, while the middle and inner octagonal rings were fabricated in the metal 8 layer. The connection structures between the outer octagonal ring and the middle octagonal ring were realized in metal 8. The metal 3 layer was used to fabricate the metallic ground plane, which could effectively prevent the waves from being exposed to the lossy substrate because metal 1 and metal 2 were used for the electronics routing of the PTAT sensor. The resistance of the polysilicon resistor was 100 Ω for impedance matching and transmission lines were formed from the octagonal ring antenna down to the resistor. A grounded wall composed of metal layers and vias layers around the antenna was applied to trap EM energy and prevent external interference. Besides, metal layers and inter-metal dielectric regions were modeled as aluminum and SiO_2, respectively, and they were fixed by the CMOS process. In addition, sample B with only a ground plane in the metal 3 layer was also simulated for verifying the frequency-selective absorption of sample A.

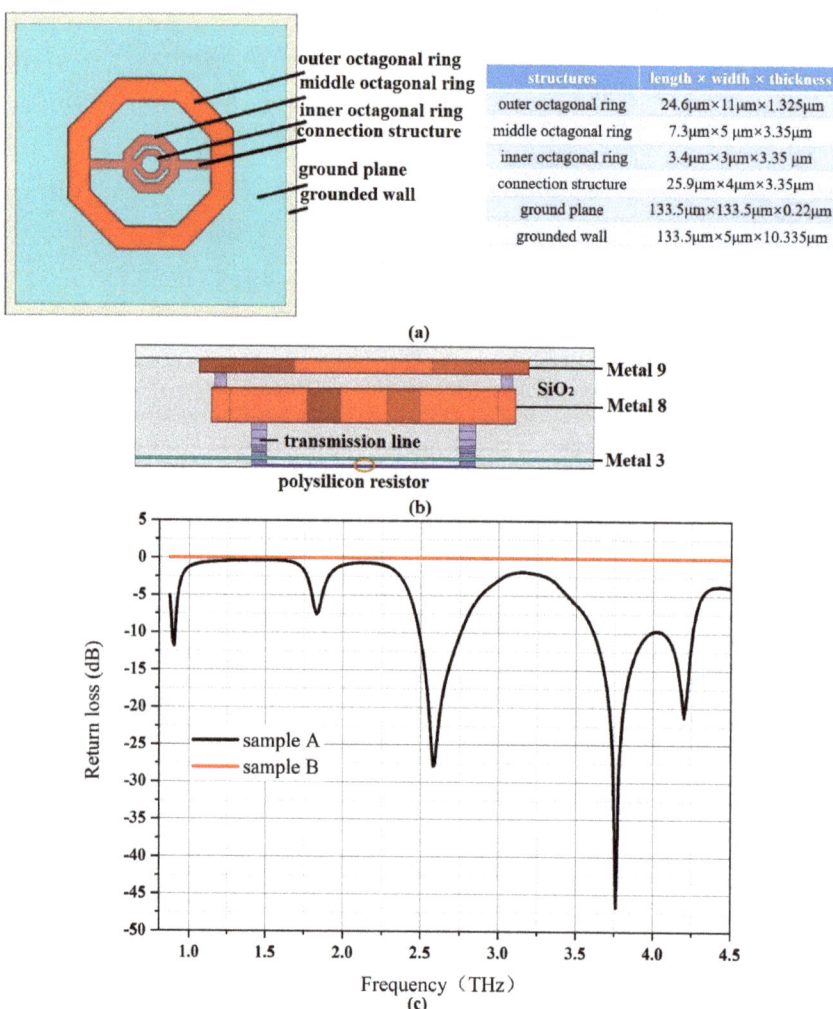

Figure 2. Designed antenna: (a) top view; (b) side view; (c) simulated return loss.

Figure 2c shows the simulated return loss curves, where sample A could resonate at 0.91 THz, 2.58 THz, and 4.2 THz, while sample B does not have frequency−selective characteristics. Within the observation frequency range, three octagonal rings radiate 0.91 THz waves, 2.58 THz waves, and 4.2 THz waves, while the second-order mode and fourth-order mode of the outer octagonal ring correspond to radiating 1.82 THz waves and 3.74 THz waves, respectively. Although sample A could radiate THz waves of five frequencies, it is still considered that a triple-band THz antenna is obtained instead of a five-band antenna. Besides, the fundamental modes of three octagonal rings are still applied to radiate 0.91 THz waves, 2.58 THz waves, and 4.2 THz waves, instead of only constructing an outer octagonal ring to obtain a THz antenna operating in multiple bands through its fundamental modes and higher modes. This is because, compared with fundamental modes, higher order modes of the antenna are unstable, and have less energy and greater loss. In addition, as the antenna operates with higher order modes, the directional pattern usually has a large domain change because of the multi-periodic current distribution. Therefore, the application of higher order modes of the antenna is generally not recommended. As sample A obtains simulated gains of 3.9 dBi, 4.24 dBi, and 3.13 dBi in the z-axis direction with simulated radiation efficiencies of 63.5%, 82.6%, and 83.4% at 0.91 THz, 2.58 THz, and 4.2 THz, respectively, the receiving efficiencies of sample A towards 0.91 THz waves, 2.58 THz waves, and 4.2 THz waves are 63.5%, 82.6%, and 83.4%, respectively.

2.2. Design of the PTAT Sensor

Bipolar junction transistors (BJTs) have become attractive temperature sensing elements because of their lower cost, better stability, higher temperature sensitivity, lower power consumption, and better process compatibility and predictability [32]. Besides, it was recognized that, if two BJTs with the same emitter area operated at different current densities or two BJTs with proportional emitter areas operated at the same current density, then their emitter–base difference voltage is proportional to the absolute temperature [33]. Therefore, PTAT sensors that convert the temperature increment into an output voltage variation proportionally constitute a type of circuit structure that relies on BJTs' temperature characteristics [34]. In order to obtain detectors with higher responsivity, there is an effective approach to adopt an optimized PTAT sensor made up of four modules with enhanced temperature sensitivity. Previous research has proposed a sensitive PTAT sensor with an increased temperature sensitivity of 10.31 mV/°C at 25 °C, as shown in Figure 3 [27]. The starting circuit is used to ensure the sensor quickly enters the normal operation state at the moment of supplying power. The PTAT current I_{PTAT} could be generated by the PTAT core circuit, and the complementary to absolute temperature (CTAT) current I_{CTAT} could be obtained from the CTAT current generation circuit. I_{PTAT} and I_{CTAT} are differentiated in a proper proportion to obtain a PTAT current with an enhanced positive temperature coefficient in the output circuit and, finally, a PTAT voltage is generated.

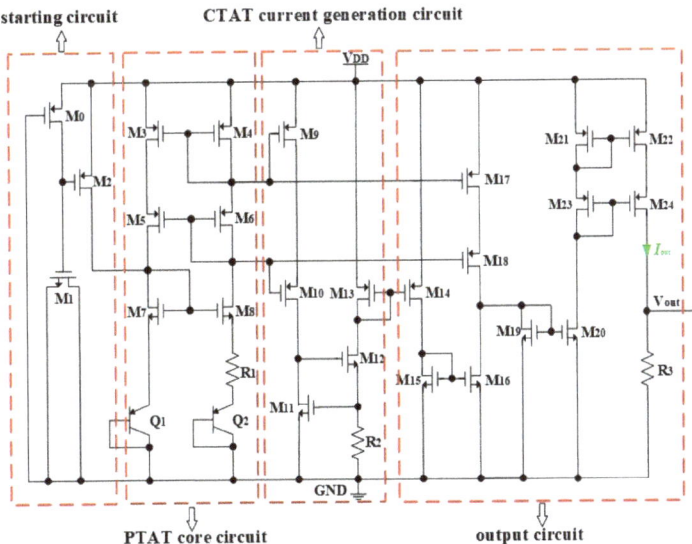

Figure 3. Schematic diagram of the optimized PTAT sensor.

2.3. Co-Design of Antenna and PTAT Sensor

As 0.91 THz waves are incident normally, under the action of electric field horizontally to the right along xoy plane, the right side of the outer octagonal ring would gather a large amount of positively induced charge, while the left side would gather equal amounts of negatively induced charge. By this means, the outer octagonal ring generates an induced electromotive force (EMF) and an induced electric field, which is opposite to the incident electric field of 0.91 THz waves. As the circumference of the outer octagonal ring is about a dielectric wavelength of 0.91 THz waves, accumulated anisotropic charges cause an induced current distributed in full wave among the outer octagonal ring, and the current finally flows to the resistor through feeding structures. Similarly, as 2.58 THz waves or 4.2 THz waves are incident normally, the middle octagonal ring or inner octagonal ring would generate an induced electric field and an induced current, which also flows to the resistor. It could be concluded that octagonal rings and the resistor constitute heat sources. Therefore, the temperature distribution of the antenna is obtained using the EM field frequency domain module and solid heat transfer module in COMSOL tools. The lumped port and boundary conditions are set to make sure that the antenna could frequency-selective receive THz waves, and incident power at the excitation port is set according to the receiving efficiency enabling the antenna to receive EM energy of 0.1 mW. Besides, the antenna and resistor are set as heat sources, and the initial temperature of the model is set to 25 °C. The temperature distribution of the antenna is obtained by frequency domain research and steady state research.

In order to clarify the layout of temperature-sensing elements and determine whether all temperature-sensing elements could be distributed within the same raised temperature distribution area, the raised temperature distribution of the antenna needs to be simulated. As the antenna receives 0.91 THz waves, the resistor at the termination of the antenna becomes the main heat source owing to ohmic loss, showing a strong temperature distribution in a certain area around the resistor. According to the antenna size, the certain area could be approximately considered as 30 μm × 30 μm, and this means that the resistor generated a strong, uniform, and raised temperature distribution in this area, as shown in Figure 4.

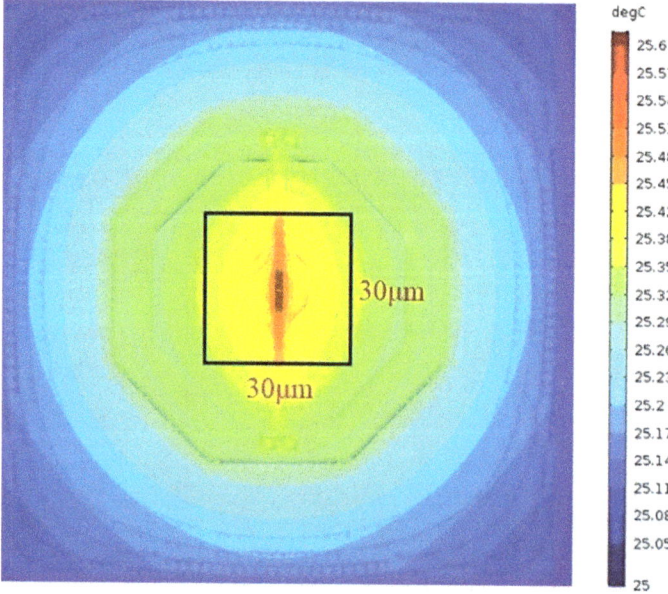

Figure 4. Temperature distribution of the proposed detector.

Then, 2.58 THz waves or 4.2 THz waves with the same EM energy of 0.1 mW were incident on the antenna, so the resistor that acts as the main heat source also generated a strong and uniform temperature in this area. Based on this, two BJTs of the PTAT sensor are the main temperature-sensing elements, and the transistor M_{11} working in the subthreshold region could also sense the raised temperature. Thus, two BJTs and M_{11} should be distributed within this area, which is centered around the resistor with an area of 30 μm × 30 μm. In this way, it not only benefits sensing the raised temperature effectively, but also could ensure that all temperature-sensing elements sense the same temperature. In addition, two BJTs are composed of a single transistor unit and seven parallel transistor units with an emitter area of 5 μm × 5 μm and an overall size of 10 μm × 10 μm. Besides, the overall size of M_{11} is 3 μm × 25 μm. It could be seen that, as two BJTs are closely arranged in a "square" shape, and there is an empty position of one transistor unit in the center to construct a polysilicon resistor, BJTs could be centrally arranged around the resistor. At the same time, when transistor M_{11} is distributed near two BJTs, the layout area formed by temperature-sensing elements is basically consistent with the temperature distribution generated by the antenna.

3. Experiment and Discussions

A die micrograph of the designed detector (DUT 1) and a device with only a PTAT sensor (DUT 2) is shown in Figure 5. DUT 1 and DUT 2 occupy chip areas of 164 μm × 214 μm and 141 μm × 193 μm, respectively. Various figures of merit, such as responsivity, noise equivalent power (NEP), and thermal time constant, are adopted to evaluate the performances of detectors.

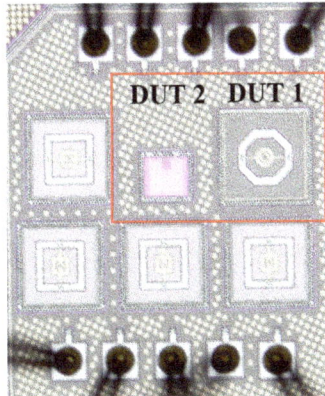

Figure 5. Die micrograph of thermal detectors.

3.1. Responsivity

The responsivity (R_v) was determined by the output voltage difference ΔV_{out} and the incident power P_{in}. ΔV_{out} was obtained by monitoring output voltages of detectors while THz waves were on and off. P_{in} was calculated by the effective area A_{eff} of the antenna and incident power density J_{in}, which was obtained by the total power of the THz beam and the area of the focused beam. R_v and A_{eff} could be expressed as follows [35]:

$$R_v = \frac{\Delta V_{out}}{P_{in}} = \frac{\Delta V_{out}}{J_{in} \cdot A_{eff}} \quad (1)$$

$$A_{eff} = \frac{G \cdot \lambda^2}{4\pi} \quad (2)$$

where G is the gain and λ is the EM wavelength in free space.

The block diagram of the output voltage measurement setup at 0.91 THz is shown in Figure 6a. The backward-wave oscillator (BWO) radiated THz waves with an average power of about 125 μW at 0.91 THz. Adjusting the height and position of two parabolic optical mirrors, as the THz beam was reflected by the first parabolic mirror, almost all THz waves were collimated and incident to the parallel second parabolic mirror. Then, after reflection and focusing, a THz beam with a diameter of about 1.15 mm was obtained. Detectors mounted on a three-dimensional stage were positioned at the focus point of the THz beam. A mechanical chopper with a minimum chopping frequency of 2 Hz modulated the THz beam, and a chopper controller modulated the chopper and the lock-in amplifier synchronously. The detector was biased at 2.5 V using a DC voltage source, and the output voltage was measured by a lock-in amplifier. As shown in Figure 6b, DUT 1 achieved the highest responsivity of 32.6 V/W as it resonated at 0.91 THz with a chopping frequency of 2 Hz. The responsivity at 0.91 THz was higher than the responsivity at both sides of 0.91 THz for the following reasons. On the one hand, the EM modeling of the antenna fully considered the design rules and implementation methods of the process in order to ensure that the antenna model was highly consistent with the layout. On the other hand, the circumference of the outer octagonal ring constructed by the metal 9 layer determined the operation frequency, and the processing tolerance of the metal 9 layer was ±0.135 μm. However, even with the maximum processing tolerance, the receiving of 0.91 THz waves was still higher than the receiving of THz waves on both sides of 0.91 THz. Furthermore, the responsivity of DUT 2 was close to zero as there was no significant voltage variation while THz waves were on or off.

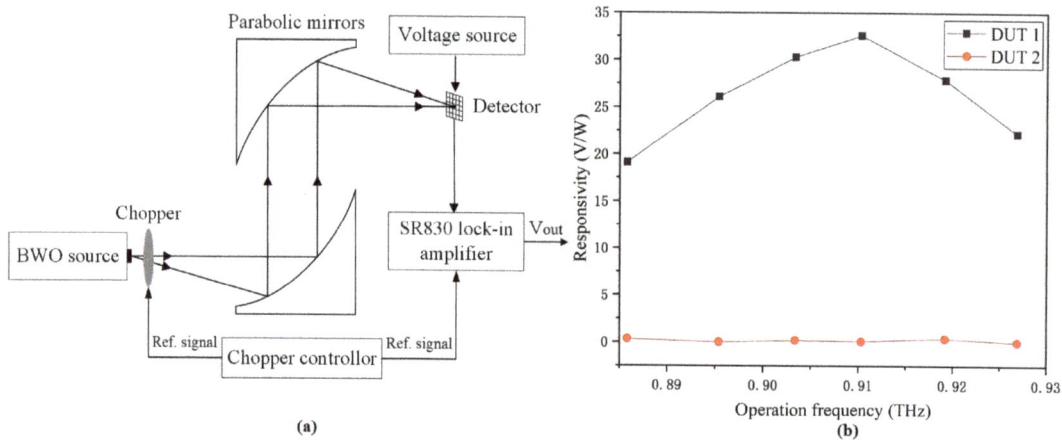

Figure 6. (**a**) Output voltage measurement setup as detectors operate at 0.91 THz. (**b**) Responsivities.

Figure 7a shows the output voltage measurement setup at 2.58 THz. A quantum cascade laser (QCL) radiated 2.58 THz waves with a peak power of 60 mW and a duty cycle of 4%. The 2.58 THz beam with a diameter of about 500 μm was collimated and focused using two parabolic optical mirrors. The detector was biased at 2.5 V and the output voltage was measured by the SR830 lock-in amplifier. The responsivity as a function of modulation frequency was acquired by modulating the QCL and the amplifier simultaneously using a signal generator. Figure 7b shows the measured responsivities of DUT 1 and DUT 2 versus modulation frequencies. Because the duration of THz waves on detectors increased as the modulation frequency decreased, the output voltage and responsivity of detectors gradually increased until the output was saturated. DUT 1 almost reached saturation at the modulation frequency of 0.5 Hz with a responsivity of 43.2 V/W, while DUT 2 showed responsivities of near zero.

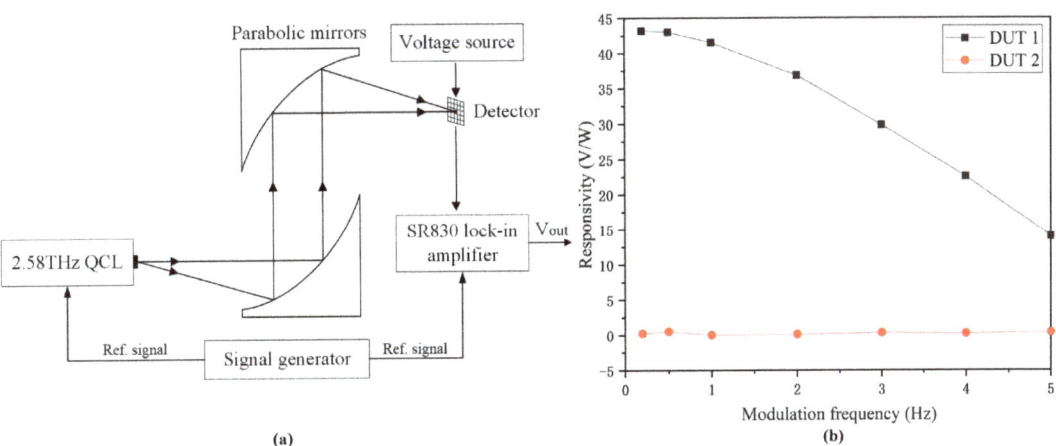

Figure 7. (**a**) Output voltage measurement setup as detectors operate at 2.58 THz. (**b**) Responsivities.

Figure 8a shows the output voltage measurement setup at 4.2 THz. A QCL provided 4.2 THz waves with an average power of about 0.5 mW and a duty cycle of 40%. The THz beam with an average power of about 0.15 mW and a diameter of about 209 μm was incident on the chip because of the strong absorption of 4.2 THz waves. The SR830 lock-in amplifier outputs synchronous modulation signals to the QCL and the output signal of the detector simultaneously. The responsivity as a function of modulation frequency from 0.3 Hz to 3 Hz was obtained from the lock-in amplifier as the detector was biased at 2.5 V. Figure 8b shows the measured responsivities of DUT 1 and DUT 2. DUT 1 almost reached saturation at the modulation frequency of 0.5 Hz with a responsivity of 40 V/W, while DUT 2 showed responsivities of near zero.

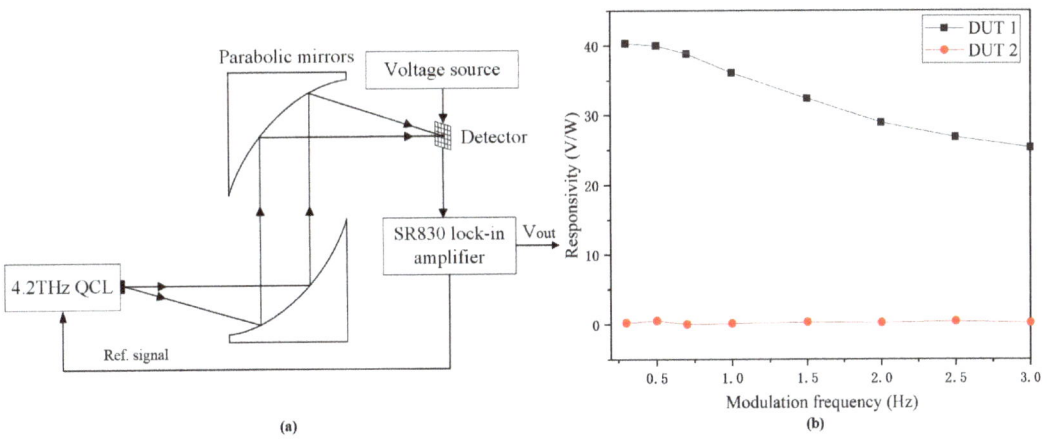

Figure 8. (**a**) Output voltage measurement setup as detectors operate at 4.2 THz. (**b**) Responsivities.

3.2. NEP

NEP was equal to the noise spectral density (NSD) divided by the responsivity. In addition, the NSD was obtained by measuring noise voltages using a dynamic signal analyzer as detectors were biased at 2.5 V [36]. As the proposed detector operated at 0.91 THz with a modulation frequency of 2 Hz, it obtained an NEP of 1.28 μW/Hz$^{0.5}$. Besides, as the detector operated at 2.58 THz and 4.2 THz, it almost reached saturation at a modulation frequency of 0.5 Hz, thus corresponding NEPs were 2.19 µW/Hz$^{0.5}$ and 2.37 µW/Hz$^{0.5}$, respectively.

3.3. Thermal Time Constant

The thermal time constant represented the duration for the temperature change and output signals rising from 0 to 63.2% of the steady-state values, deriving from the responsivity versus modulation frequency [36]. The thermal time constant values of the proposed detector operating at 2.58 THz and 4.2 THz were extracted as 298 ms and 330 ms, respectively.

3.4. Performance Summary and Comparison

Although multiband thermal detectors show lower responsivity and higher NEP than multiband semiconductor detectors, they exactly constitute compact multiband structures, allow a wide spectrum detection, and avoid multiband impedance matching. Therefore, the proposed triple-band THz thermal detector and several published thermal detectors are summarized and compared in Table 1. A quad-band thermal detector consisting of an antenna and an NMOS sensor is designed to operate at 0.546 THz, 0.688 THz, 0.78 THz, and 0.912 THz [26]. It obtains better characteristic results, but operation at higher frequencies is preferable, thus THz thermal detectors operating above 1 THz have been proposed. Thermal detectors operating at three single frequencies of 1 THz, 2.9 THz, and 28.3 THz are composed of three discrete detectors with worse characteristic results [24]. In addition, compared with triple-band detectors in [27,28], the proposed detector applied an octagonal ring antenna as an alternative absorbing structure, showing a slight difference in performance due to the processing error or measurement error. Besides, the proposed detector, which was measured at three operation frequencies, also highlighted the concept of collaborative designs and detailed analysis. Although performance measurements of the proposed detector were finished at three operation frequencies, we would like to obtain traces of the output voltage from an oscilloscope or source meter at a certain chopping frequency for the purpose of visualizing in-time performances greatly in the future. Furthermore, in future work, we also prefer to compare performances at different ambient temperatures in order to further estimate the best possible performances.

Table 1. Performance summary and comparison.

Ref. No	Frequency (THz)	Structure	Technology	Rv (V/W)	NEP (W/$\sqrt{Hz}$)
[26]	0.546, 0.688, 0.78, 0.912	Antenna + NMOS sensor	0.18 µm CMOS	5.5 k, 5.3 k, 3 k, 5 k	0.94 p, 0.98 p, 1.72 p, 1.03 p
[24]	1, 2.9, 28.3	Dipole antenna + PTAT sensor	0.18 µm CMOS	18 *, 18.9, 18.6 *	1.7 µ *
[27]	0.91, 2.58, 4.3	Metamaterial absorber + PTAT sensor	55 nm CMOS	33.4, 47.9, 61.11 *	1.49, 1.88, 1.31 * µ
[28]	0.91, 2.58, 4.3	Loop antenna + PTAT sensor	55 nm CMOS	29.2, 46.5, 47.6 *	1.57, 1.26, 3.29 * µ
This work	0.91, 2.58, 4.2	Octagonal ring antenna + PTAT sensor	55 nm CMOS	32.6, 43.2, 40	1.28, 2.19, 2.37 µ

* based on simulation results.

4. Conclusions

To conclude, this paper presents the design, simulation, and experimental validation of a room-temperature monolithic resonant triple-band THz thermal detector, which was fully implemented in a CMOS process, allowing reduced fabrication complexity and lower production cost. The detector was composed of a compact triple-band octagonal ring antenna loaded with a polysilicon resistor and a sensitive PTAT sensor. The responsivity, noise equivalent power, and thermal time constant of the detector were experimentally assessed at 0.91 THz, 2.58 THz, and 4.2 THz, showing relatively better measurement results. The detector also has natural scalability to focal plane arrays, demonstrating significant advances in developing compact, room-temperature, low-cost, and mass-production multiband THz detection systems.

Author Contributions: Conceptualization, X.W., T.-P.L., S.-X.Y. and J.W.; methodology, X.W. and S.-X.Y.; software, X.W. and S.-X.Y.; validation, X.W., T.-P.L., S.-X.Y. and J.W.; formal analysis, J.W.; investigation, X.W., T.-P.L., S.-X.Y. and J.W.; resources, T.-P.L. and J.W.; data curation, X.W. and S.-X.Y.; writing—original draft preparation, X.W. and S.-X.Y.; writing—review and editing, S.-X.Y. and J.W.; visualization, X.W. and S.-X.Y.; supervision, S.-X.Y. and J.W.; project administration, S.-X.Y. and J.W.; funding acquisition, T.-P.L. and J.W. All authors have read and agreed to the published version of the manuscript.

Funding: This work was supported by the State Key Laboratory of Complex Electromagnetic Environment Effects on Electronics and Information System (No. CEMEE2022G0201) and the China Postdoctoral Science Foundation (Grant No. 2020M680883).

Data Availability Statement: Not applicable.

Conflicts of Interest: The authors declare no conflict of interest.

References

1. Malhotra, I.; Jha, K.R.; Singh, G. Terahertz antenna technology for imaging applications: A technical review. *Int. J. Microw. Wirel. Technol.* **2018**, *10*, 271–290. [CrossRef]
2. Yu, C.; Fan, S.; Sun, Y.; Pickwell-MacPherson, E. The potential of terahertz imaging for cancer diagnosis: A review of investigations to date. *Quant. Imag. Med. Surg.* **2012**, *2*, 33–45.
3. Yang, X.; Zhao, X.; Yang, K.; Liu, Y.P.; Liu, Y.; Fu, W.; Luo, Y. Biomedical applications of terahertz spectroscopy and imaging. *Trends Biotechnol.* **2016**, *34*, 810–824. [CrossRef]
4. Kemp, M.C. Explosives detection by terahertz spectroscopy-a bridge too far? *IEEE Trans. Terahertz Sci. Technol.* **2011**, *1*, 282–292. [CrossRef]
5. Danciu, M.; Alexa, T.; Stefanescu, C.; Dodi, G.; Tamba, B.I.; Mihai, C.T.; Stanciu, G.D.; Luca, A.; Spiridon, I.A.; Ungureanu, L.B.; et al. Terahertz Spectroscopy and Imaging: A Cutting-Edge Method for Diagnosing Digestive Cancers. *Materials* **2019**, *12*, 1519. [CrossRef]
6. Takida, Y.; Nawata, K.; Minamide, H. Security screening system based on terahertz-wave spectroscopic gas detection. *Opt. Express* **2021**, *29*, 2529–2537. [CrossRef]
7. Karaliūnas, M.; Nasser, K.E.; Urbanowicz, A.; Kašalynas, I.; Bražinskienė, D.; Asadauskas, S.; Valušis, G. Non-destructive inspection of food and technical oils by terahertz spectroscopy. *Sci. Rep.* **2018**, *8*, 18025. [CrossRef]
8. Yi, C.; Kim, D.; Solanki, S.; Kwon, J.H.; Kim, M.; Jeon, S.; Ko, Y.C.; Lee, I. Design and Performance Analysis of THz Wireless Communication Systems for Chip-to-Chip and Personal Area Networks Applications. *IEEE J. Sel. Area. Commun.* **2021**, *39*, 1785–1796. [CrossRef]
9. Xu, H.; Lu, H.; Wang, Z.; Liu, J.; Wang, W. The System Design and Preliminary Tests of the THz Atmospheric Limb Sounder (TALIS). *IEEE Trans. Instrum. Meas.* **2022**, *71*, 1–12. [CrossRef]
10. Bower, G.C.; Dexter, J.; Asada, K.; Brinkerink, C.D.; Falcke, H.; Ho, P.; Inoue, M.; Markoff, S.; Marrone, D.P.; Matsushita, S.; et al. ALMA observations of the terahertz spectrum of sagittarius A*. *Astrophys. J.* **2019**, *881*, L2. [CrossRef]
11. Zhou, D.; Hou, L.; Xie, W.; Zang, W.; Lu, B.; Chen, J.; Wu, P. Practical dual-band terahertz imaging system. *Appl. Opt.* **2017**, *56*, 3148–3154. [CrossRef]
12. Dabironezare, S.O.; Hassel, J.; Gandini, E.; Gronberg, L.; Sipola, H.; Vesterinen, V.; Llombart, N. A dual-band focal plane array of kinetic inductance bolometers based on frequency-selective absorbers. *IEEE Trans. Terahertz Sci. Technol.* **2018**, *8*, 746–756. [CrossRef]
13. Cheng, Q.; Paradis, S.; Bui, T.; Almasri, M. Design of Dual-Band Uncooled Infrared Microbolometer. *IEEE Sens. J.* **2011**, *11*, 167–175. [CrossRef]
14. Statnikov, K.; Grzyb, J.; Heinemann, B.; Pfeiffer, U.R. 160-GHz to 1-THz multi-color active imaging with a lens-coupled SiGe HBT chip-set. *IEEE Trans. Microw. Theory Tech.* **2015**, *63*, 520–532. [CrossRef]
15. Nguyen, T.D.; Hong, J.P. A High Fundamental Frequency Sub-THz CMOS Oscillator With a Capacitive Load Reduction Circuit. *IEEE Trans. Microw. Theory Tech.* **2020**, *68*, 2655–2667. [CrossRef]
16. Huang, R.; Ji, Y.; Liao, Y.; Peng, J.; Wang, K.; Xu, Y.; Yan, F. Dual-frequency CMOS terahertz detector with silicon-based plasmonic antenna. *Opt. Express* **2019**, *27*, 23250–23261. [CrossRef]
17. Xu, L.J.; Guan, J.N.; Bai, X.; Li, Q.; Mao, H.P. A novel CMOS multi-band THz detector with embedded ring antenna. *J. Infrared Millim. Terahertz Waves* **2017**, *38*, 1189–1205. [CrossRef]
18. Khatib, M.; Perenzoni, M. Response optimization of antenna-coupled FET detectors for 0.85-to-1-THz imaging. *IEEE Microw. Wirel. Compon. Lett.* **2018**, *28*, 903–905. [CrossRef]

19. Boppel, S.; Lisauskas, A.; Mundt, M.; Seliuta, D.; Minkevičius, L.; Kašalynas, I.; Valušis, G.; Mittendorff, M.; Winnerl, S.; Krozer, V.; et al. CMOS integrated antenna-coupled field-effect transistors for the detection of radiation from 0.2 to 4.3THz. *IEEE Trans. Microw. Theory Tech.* **2012**, *60*, 3834–3843. [CrossRef]
20. Bauer, M.; Venckevicius, R.; Kasalynas, I.; Boppel, S.; Mundt, M.; Minkevicius, L.; Lisauskas, A.; Valusis, G.; Krozer, V.; Roskos, H.G. Antenna-coupled field-effect transistors for multi-spectral terahertz imaging up to 4.25THz. *Opt. Express* **2014**, *22*, 19235–19241. [CrossRef]
21. Hillger, P.; Grzyb, J.; Jain, R.; Pfeiffer, U.R. Terahertz Imaging and Sensing Applications with Silicon-Based Technologies. *IEEE Trans. Terahertz Sci. Technol.* **2019**, *9*, 1–19. [CrossRef]
22. Grossman, E.; Dietlein, C.; Ala-Laurinaho, J.; Leivo, M.; Gronberg, L.; Gronholm, M.; Lappalainen, P.; Rautiainen, A.; Tamminen, A.; Luukanen, A. Passive terahertz camera for standoff security screening. *Appl. Opt.* **2010**, *49*, E106–E120. [CrossRef]
23. Nguyen, D.T.; Simoens, F.; Ouvrier-Buffet, J.; Meilhan, J.; Coutaz, J.L. Broadband THz uncooled antenna-coupled microbolometer array-electromagnetic design, simulations and measurements. *IEEE Trans. Terahertz Sci. Technol.* **2012**, *2*, 299–305. [CrossRef]
24. Chen, F.; Yang, J.; Li, Z. Modeling of an uncooled CMOS THz thermal detector with frequency-selective dipole antenna and PTAT temperature sensor. *IEEE Sens. J.* **2018**, *18*, 1483–1492. [CrossRef]
25. Kuzmin, L.S.; Sobolev, A.S.; Beiranvand, B. Wideband Double-Polarized Array of Cold-Electron Bolometers for OLIMPO Balloon Telescope. *IEEE Trans. Antennas Propagat.* **2021**, *69*, 1427–1432. [CrossRef]
26. Zhang, Y.; Zhang, S. Novel CMOS-based multi-band terahertz detector for passive imaging. *Semi. Sci. Technol.* **2022**, *37*, 055014. [CrossRef]
27. Wang, X. Uncooled CMOS Integrated Triple-Band Terahertz Thermal Detector Comprising of Metamaterial Absorber and PTAT Sensor. *IEEE Access* **2020**, *8*, 114501–114508. [CrossRef]
28. Wang, X. Monolithic resonant CMOS fully integrated triple-band THz thermal detector. *Opt. Express* **2020**, *28*, 22630–22641. [CrossRef]
29. Corcos, D.; Kaminski, N.; Shumaker, E.; Markish, O.; Elad, D.; Morf, T.; Drechsler, U.; Saha, W.T.S.; Kull, L.; Wood, K.; et al. Antenna-coupled MOSFET bolometers for uncooled THz sensing. *IEEE Trans. Terahertz Sci. Technol.* **2015**, *5*, 902–913. [CrossRef]
30. Li, S.; Gao, J.; Cao, X.; Li, W.; Zhang, Z.; Zhang, D. Wideband, thin, and polarization-insensitive perfect absorber based the double octagonal rings metamaterials and lumped resistances. *J. Appl. Phys.* **2014**, *116*, 207402. [CrossRef]
31. Golda, A.; Kos, A. Analysis and design of PTAT temperature sensor in digital CMOS VLSI circuits. In Proceedings of the International Conference Mixed Design of Integrated Circuits and System, Gdynia, Poland, 22–24 June 2006; pp. 415–420.
32. Li, Y. The Research and Design of Low Power Low Offset Digital Thermometer. Master Thesis, Hunan University, Changsha, China, 2006.
33. Razavi, B. *Design of Analog CMOS Integrated Circuits*; Xi'an Jiaotong University Press: Xi'an, China, 2003.
34. Liu, M.C.; Hsiao, S.Y. PTAT Sensor and Temperature Sensing Method Thereof. U.S. Patent US7915947B2, 29 March 2011.
35. Liu, Z.; Liu, L.; Yang, J.; Wu, N. A CMOS fully integrated 860-GHz terahertz sensor. *IEEE Trans. Terahertz Sci. Technol.* **2017**, *7*, 455–465. [CrossRef]
36. Carranza, I.E.; Grant, J.; Gough, J.; Cumming, D.R.S. Metamaterial-Based Terahertz Imaging. *IEEE Trans. THz Sci. Technol.* **2015**, *5*, 892–901. [CrossRef]

Disclaimer/Publisher's Note: The statements, opinions and data contained in all publications are solely those of the individual author(s) and contributor(s) and not of MDPI and/or the editor(s). MDPI and/or the editor(s) disclaim responsibility for any injury to people or property resulting from any ideas, methods, instructions or products referred to in the content.

Article

A G-Band Broadband Continuous Wave Traveling Wave Tube for Wireless Communications

Yuan Feng *, Xingwang Bian, Bowen Song, Ying Li, Pan Pan and Jinjun Feng

National Key Laboratory of Science and Technology on Vacuum Electronics, Beijing Vacuum Electronics Research Institute, Beijing 100015, China
* Correspondence: fengyuan_bveri@foxmail.com; Tel.: +86-010-8435-2321

Abstract: Development of a G-band broadband continuous wave (CW) traveling wave tube (TWT) for wireless communications is described in this paper. This device provides the saturation output power over 8 W and the saturation gain over 30.5 dB with a bandwidth of 27 GHz. The maximum output power is 16 W and the bandwidth of 10 W output power is 23 GHz. The 3 dB bandwidth is greater than 12.3% of f_c (center frequency). The gain ripple is less than 10 dB in band. A pencil beam of 50 mA and 20 kV is used and a transmission ratio over 93% is realized. The intercept power of the beam is less than 70 W and the TWT is conduction cooled through mounting plate and air fan, which makes the device capable of operating in continuous wave mode. A Pierce's electron gun and periodic permanent magnets are employed. Chemical vapor deposition diamond disc is used in the input and output radio frequency (RF) windows to minimize the loss and voltage standing wave ratios of the traveling wave tube. Double stages deeply depressed collector is used for improving the total efficiency of the device, which can be over 5.5% in band. The weight of the device is 2.5 kg, and the packaged size is 330 mm × 70 mm × 70 mm.

Keywords: G-band broadband amplifiers; traveling wave tubes; folded waveguide

1. Introduction

G-band electromagnetic wave provides availability for the design of terrestrial and satellite radio communication networks according to the radio regulations of International Telecommunication Union.

The European Commission Horizon 2020 ULTRAWAVE, "Ultra-capacity wireless layer beyond 100 GHz based on millimeter wave Traveling Wave Tubes", aims to exploit portions in the millimeter wave spectrum for creating a very high-capacity layer [1].

However, there are two problems. One problem is the atmospheric attenuation, which directly influences using these frequencies for long range communication [2]. The high atmosphere attenuation and the lack of enough transmission power limit the range to a few tens of meters, even by using high gain antennas [3].

Another problem is that there is a frequency range called "Terahertz Gap" between the highest frequency of microwave technology and the lowest frequency of photonic technology [4].

Vacuum electronic devices (VEDs) exhibit the advantages of high average power, high operation frequency, and high efficiency. Traveling wave tubes (TWTs), one of the most widely used VEDs, exhibit the incomparable advantage of wideband, which is much higher than that of the available solid-state devices [5]. In THz regime, TWTs are the most widely used VEDs.

In recent years, great progress has been made in the development of G-band TWT. Some TWTs operating at G-band have been demonstrated [6–11], and the performances of the TWTs are shown in Table 1.

Citation: Feng, Y.; Bian, X.; Song, B.; Li, Y.; Pan, P.; Feng, J. A G-Band Broadband Continuous Wave Traveling Wave Tube for Wireless Communications. *Micromachines* 2022, 13, 1635. https://doi.org/10.3390/mi13101635

Academic Editors: Lu Zhang, Xiaodan Pang and Prakash Pitchappa

Received: 5 September 2022
Accepted: 26 September 2022
Published: 29 September 2022

Publisher's Note: MDPI stays neutral with regard to jurisdictional claims in published maps and institutional affiliations.

Copyright: © 2022 by the authors. Licensee MDPI, Basel, Switzerland. This article is an open access article distributed under the terms and conditions of the Creative Commons Attribution (CC BY) license (https://creativecommons.org/licenses/by/4.0/).

Table 1. Some G-band TWTs performances in recent years.

No.	Output Power	Bandwidth	Gain	Operational Duty Cycles	Organization
1	50 W	2.4 GHz	—	50%	Northrop Grumman
2	50 W	3.6 GHz	35 dB	5%	BVERI
3	11 W	14 GHz	27 dB	—	UC Davis
4	9 W	10 GHz	25 dB	CW	China Academy of Engineering Physics
5	14.1 W	7 GHz	30.7 dB	CW	China Academy of Engineering Physics
6	15 W	7.6 GHz	32 dB	CW	BVERI
7	8 W (3 dB)	27 GHz (12.3% f_c)	30.5 dB	CW	This TWT
	10 W	23 GHz			

The bandwidth of most of the TWTs have narrower bandwidth ($\leq$10 GHz), and half of them cannot operate at continue wave (CW) mode. These performances of these devices limited the applications in wireless communications, which require wider bandwidth (3 dB bandwidth $\geq$10%), higher operational duty cycles (100%), higher total efficiency ($\geq$5%), higher gain ($\geq$30 dB), and lower gain ripple ($\leq$10 dB in band).

In order to solve the problems of the above G-band TWT and fit the requirements of terahertz communication applications, a G-band wideband continuous wave TWT is designed by Beijing Vacuum Electronics Research Institute (BVERI) and described in this article. The device is developed according to the following design routes, which is different from normal.

(1) In order to realize wideband beam–wave interaction, phase shift beyond 540° is used as the working points, where the coupling impedance is low, but the dispersion is flat.
(2) The highest frequency (230 GHz) in band is used as the reference frequency instead of the center frequency, which is beneficial to the optimization of structure parameters and electrical parameters to reduce the gain ripple.
(3) A double stages deeply depressed collector is used for improving the total efficiency of the TWT.

By the above design routes, a G-band TWT with a continuous wave output power of 8 W and a gain of 30.5 dB with 27 GHz bandwidth is realized. The maximum output power is 16 W and the bandwidth of 10 W output power is 23 GHz. The 3 dB bandwidth is greater than 12.3% of f_c (center frequency). The gain ripple is less than 10 dB in band.

2. TWT Design and Simulation

The TWT primarily contains five parts: electron gun, focusing system, radio frequency (RF) circuit, RF windows, and collector. The building blocks of the TWT are shown in Figure 1.

A Pierce's electron gun is used to produce an electron beam with a current of 20 kV and 50 mA. The type of the cathode is M-type and the cathode loading is 5 A/cm^2. A focus electrode modulates the beam providing a duty cycle from 0.1% to continuous wave. The double anodes adjust the beam current and transmission. Opera 3D is used to design the electron gun, and the simulation result is shown in Figure 2. The designed beam voltage and current of the TWT are 20 kV and 50 mA with a beam radius of 0.06 mm and a beam-shot of 10 mm, respectively.

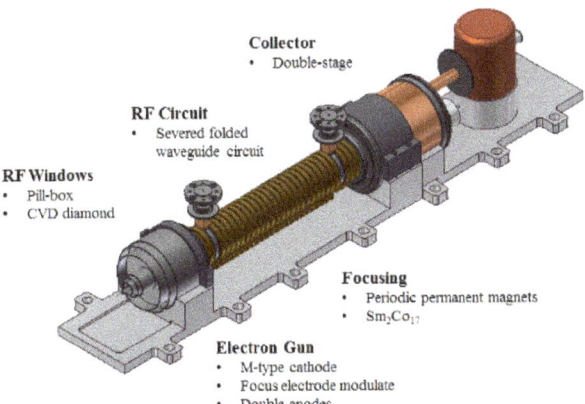

Figure 1. Building blocks of the G-band TWT.

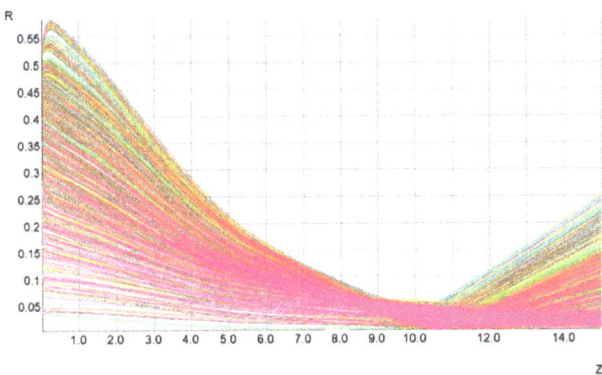

Figure 2. Simulation result of the electron gun.

Sm_2Co_{17} periodic permanent magnet (PPM) is used as the focusing system. The samarium cobalt magnets produce an on-axis B_z whose peak value is 0.5 T. The system is simulated by opera 3D. Figure 3 shows the simulation model. According to the simulation results, a high beam transmission ratio of 99% through the small diameter beam tunnel in the slow wave circuit is essential.

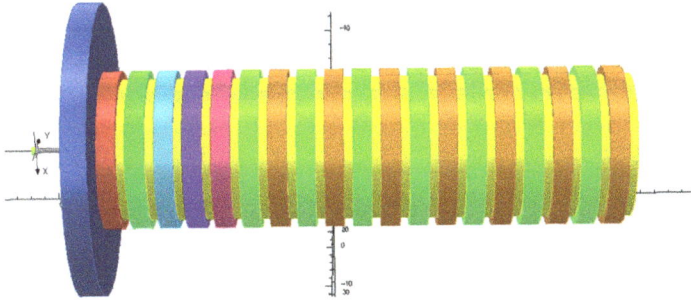

Figure 3. Model of Sm2Co17 periodic permanent magnet focusing system.

A folded waveguide is employed as the slow-wave structure in the TWT. It is fabricated with CNC-machining. The material chosen was oxygen free copper. The fabricated circular is shown in Figure 4.

Figure 4. The fabricated circular of the folded waveguide slow-wave structure.

Figure 5 shows the dispersion curves of a traditional folded waveguide circuit. For traditional fold waveguide slow-wave structure, $n = -1$ forward branch of the dispersion curve was used for the circuit, which operated as usual.

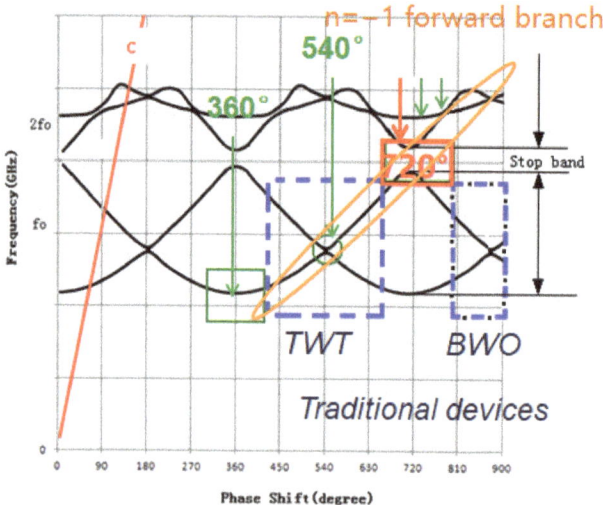

Figure 5. Dispersion curves of a traditional folded waveguide circuit.

For the fully symmetric folded waveguide slow-wave structure, there is no stop band at phase shift of 540° in theory. However, due to the machining, the actual slow wave structure will have a certain asymmetry, which may result in a stopband at the phase shift of 540° [12]. This stopband may affect the matching characteristics at the frequency, which may depress the output power at this frequency. In addition, it may result in the undesired 3π oscillations. Therefore, phase shift of 540° is usually avoided to fall into the operating frequency band when designing the folded waveguide slow-wave structure.

According to the calculation formula of coupling impedance

$$K_c = \frac{|E_z|^2}{2\beta P}$$

for folded waveguide slow-wave structures, the closer to the cutoff frequency, the stronger coupling impedance is. Here, E_z is the axial component of the electric field, β is the phase constant of the electromagnetic wave, and P is the power flowing. Therefore, in order to obtain greater power and higher interaction efficiency, the operating point is usually selected within 540°.

However, as shown in Figure 5, the phase shift within 540° means that the dispersion is stronger, which directly affects the operating bandwidth of the TWT. As the dispersion intensity of the folded waveguide slow-wave structure is positively correlated with the proportion of the beam injection channel size to the waveguide, due to the limitations of the performance of the current focusing system, it is difficult to further reduce the size of the beam injection channel in the THz band.

One way to achieve broadband performance is selecting the operating point above 540°, which is not usually used for its lower coupling impedance. The dispersion is more flat, which is more conducive to the synchronization and interaction between beam and wave. In addition, as the operating point moves away from the cutoff frequency, high frequency loss can be reduced. The saved power from high frequency loss can be stored in the beam and recovered by the deeply depressed collector with an efficiency of more than 90%.

At the same time, the decrease of electronic efficiency can inhibit the dynamic defocus of the beam, which is beneficial to improve the dynamic flow rate.

By the above design routes, the G-band broadband folded waveguide slow-wave structure is designed, and its cold characteristics are calculated by CST Microwave Studio. The simulation results are shown in Figures 6–9.

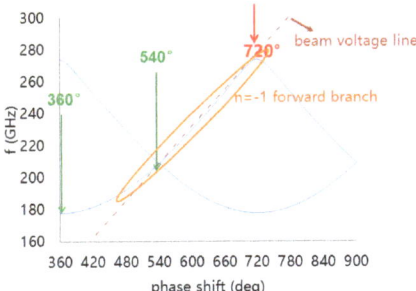

Figure 6. Dispersion curves of the folded waveguide circuit.

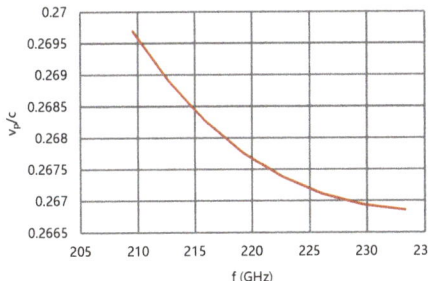

Figure 7. Normalized phase velocity of the folded waveguide circuit.

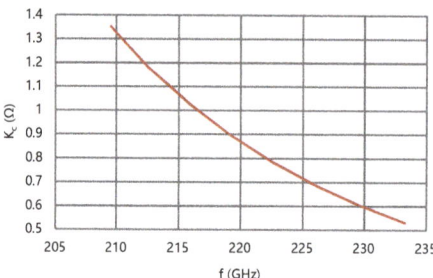

Figure 8. Coupling impedance of the folded waveguide circuit.

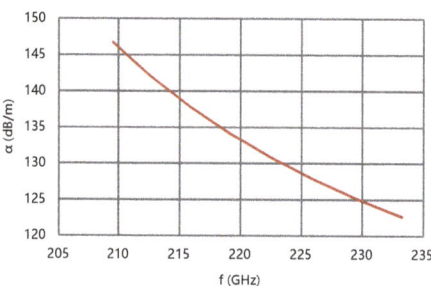

Figure 9. Attenuation coefficient of the folded waveguide circuit.

Considering the effects of dispersion, coupling impedance, and attenuation, the working point is selected between 540° to 630°. The beam line almost coincides with all frequency points from 208 GHz to 233 GHz, which ensures excellent beam–wave synchronization in the band and allows a wideband beam–wave interaction.

According to the results of Figures 7 and 8, the difference of phase velocity is less than 1% of v_{pc} (center) and the coupling impedance of the folded waveguide circuit is over 0.5 Ω in band. The effective conductivity of the circuit is empirically set as 2.6×10^7 S/m considering the surface roughness. The attenuation coefficient of the folded waveguide circuit is less than 150 dB/m, as shown in Figure 9.

The severed folded waveguide circuit consists of an input section and an output section, which ensures the stable operation while providing a high gain over 30 dB with low ripples.

The center frequency is usually selected as the reference frequency to find the operating voltage when designing TWTs. The method is applicable when designing low frequency TWTs or THz narrow band TWTs, as their variation of in-band coupling impedance is small. However, for THz wideband TWT, the in-band coupling impedance varies strongly, and the coupling impedance of which at the low frequency may be more than three times that at the high frequency end, resulting in large gain ripple.

The design scheme in this paper does not follow the traditional design scheme, and the highest frequency in band has been taken as the reference frequency to determine the operation voltage. By adjusting the dispersion strength of the slow-wave structure, the beam–wave interaction performance at the low-frequency can be adjusted, which can bridge the gain ripple caused by the change of coupling impedance.

The performance of the circuit is simulated by using a large signal beam–wave interaction software microwave tube simulator suite (MTSS). The saturation output power of the circuit is over 12 W and the saturation gain is over 27.8 dB in 208–233 GHz at the designed beam voltage and current, as shown in Figures 10 and 11.

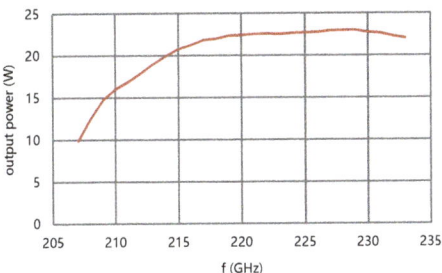

Figure 10. Saturation output power of the circuit.

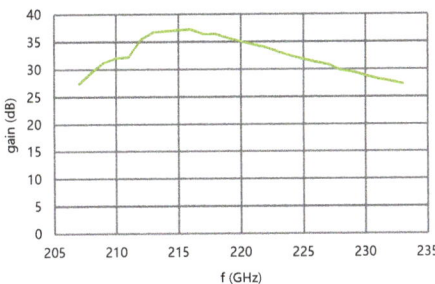

Figure 11. Saturation gains of the circuit.

In order to increase the total efficiency, a double stage depressed collector with an efficiency of over 90% is used in the TWT. The design voltage of the first stage is 17.5 kV, and the voltage of the second stage is 18.5 kV. The simulation results are shown in Figure 12 and Table 2. According to the results, the total efficiency of the TWT can be over 8% in band.

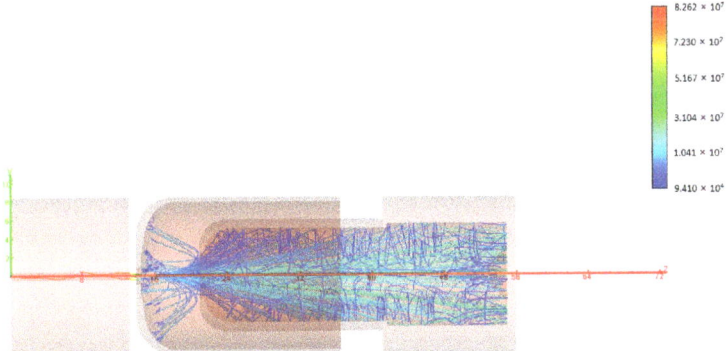

Figure 12. Electron distribution in the double stages depressed collector.

Table 2. The simulation results of different frequency.

f (GHz)	Collector Efficiency (%)	Total Efficiency (%)
210	91.73	8.03
220	91.38	9.5
230	91.99	8.15

Diamond is used as window disk material because of its small dielectric constant, small loss tangent, high thermal conductivity, and good broadband matching. Both the input and output RF windows of the TWT are pillbox windows, and the waveguide standard WR-4 is selected according to the operating frequency. CST Microwave Studio was used to optimize the S-parameters of the window. The measured S_{21} of a typical RF window is about −1 dB and the S_{11} is lower than −10 dB in 200–240 GHz, as shown in Figure 13.

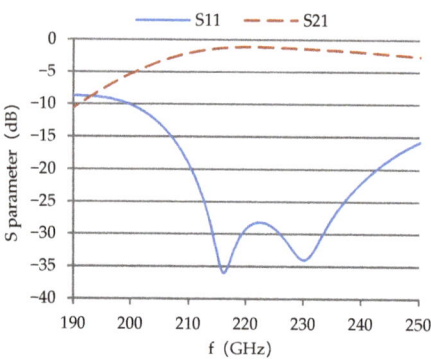

Figure 13. Measured S11 and S21 of a typical RF window employed in the TWT.

3. TWT Performance

The block diagram of the experimental setup is shown in Figure 14, and the test system is shown in Figure 15. A solid-state amplifier-multiplier chain (AMC) is used to provide the input power for the TWT. Two directional couplers are used to sample the input and output power of the TWT, respectively. The input and output power are measured with two THz power meters simultaneously.

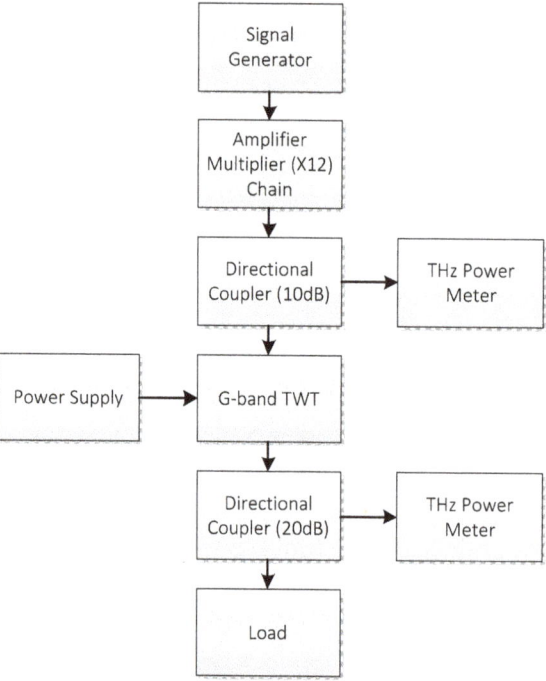

Figure 14. Block diagram of the experimental setup.

Figure 15. Test system.

The TWT operates in continuous wave mode at an optimal voltage of 20 kV and a beam current of 50 mA. The body current is 3 mA without RF and the worst body current with RF is 3.5 mA. The corresponding electron transmission ratio is over 93%. The TWT is conduction cooled through the mounting plate.

The measured output power and gain against input power for the TWT at different frequencies are shown in Figures 16 and 17. The input and output segments of the slow-wave structure have the same size in the design, and the AM (amplitude modulation) /AM was not specifically considered. We plan to use anomalous dispersion to improve linearity in the future.

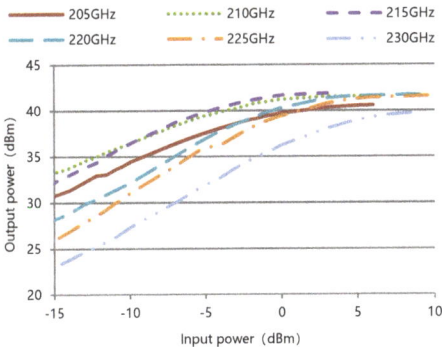

Figure 16. Measured output power against input power for the TWT at different frequencies.

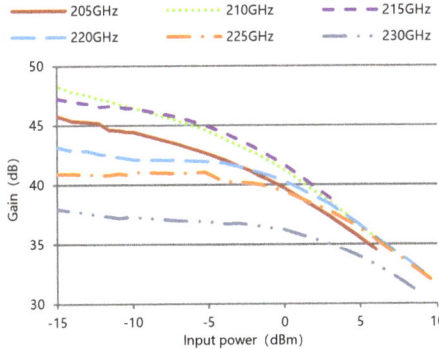

Figure 17. Measured gain against input power for the TWT at different frequencies.

The measured saturation output power and gain of the TWT are shown in Figures 18 and 19. The saturation output power is over 8 W and the saturation gain is over 30.5 dB in 204–231 GHz. The saturation output power is over 10 W in 205–228 GHz. The maximum output power is 16 W at 218 GHz. The 3 dB bandwidth is greater than 12.3% of f_c. The gain ripple is less than 10 dB in band.

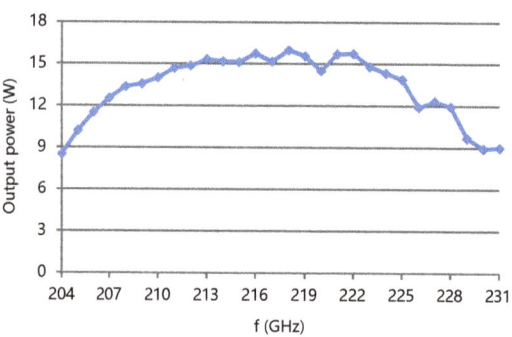

Figure 18. Measured saturation output power of the TWT.

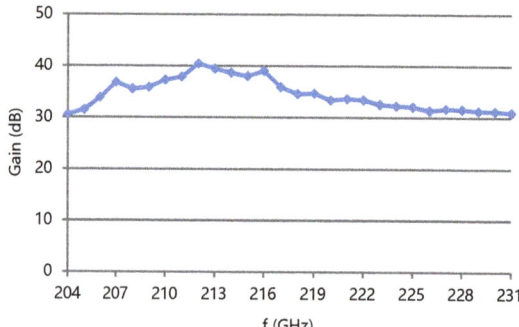

Figure 19. Measured gain of the TWT.

Comparing the measured saturation output power of the TWT with the simulation results, the output power is lower than simulated. One reason for this is that the beam current is set as 50 mA in the simulation, however, electrons of 3.5 mA are intercepted before they research the output port of the TWT, which can depress the beam–wave interaction efficiency. Another reason is that the insertion loose of the RF window is estimated as 1.5 dB in the simulation, which has been verified by the cold test of the RF window.

In addition, the maximum output power of the solid-stage source is less than 8 mW beyond 228 GHz. This is why the output power beyond 228 GHz reduced significantly.

Comparing the measured gain of the TWT with the simulation results, the gain is higher than simulated. A possible reason is that the input beam is thicker in the input part than design. This can let the average coupling impedance in the beam cross section be higher than the design. Stronger beam–wave interaction can occur, and the gain is higher.

The maximum efficiency of the TWT can research 10.2% and the total efficiency is over 5.5% in band, corresponding to a power dissipation of less than 160 W, as shown in Figure 20.

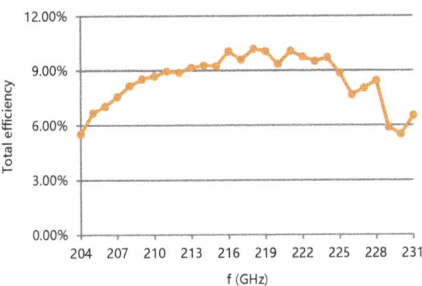

Figure 20. Measured total efficiency of the TWT.

Figure 21 is the photo of the packaged G-band TWT. The weight of the packaged TWT is 2.5 kg and the size is 330 mm × 70 mm × 70 mm, respectively.

Figure 21. The photo of the packaged G-band TWT.

4. Conclusions

This article presents the development of a G-band broadband continuous wave TWT for wireless communications based on a slow-wave structure of fold waveguide. The device provides the saturation output power over 8 W and the saturation gain over 30.5 dB with a bandwidth of 27 GHz. The maximum output power is 16 W and the bandwidth of 10 W output power is 23 GHz. The 3 dB bandwidth is greater than 12.3% of f_c. The gain ripple is less than 10 dB in band. A pencil beam of 50 mA and 20 kV is used and a transmission ratio over 93% is realized. A double stages deeply depressed collector was used for improving the total efficiency of the device, which can be over 5.5% in band. The weight of the device is 2.5 kg, and the packaged size is 330 mm × 70 mm × 70 mm.

Author Contributions: Design and simulation, Y.F., X.B.; debugging and measuring, Y.F., B.S.; machining, Y.F., Y.L.; technical support, P.P., J.F. All authors have read and agreed to the published version of the manuscript.

Funding: This research was funded by National Key Research and Development Program of China, grant number 2019YFB1803200.

Institutional Review Board Statement: Not applicable.

Informed Consent Statement: Not applicable.

Data Availability Statement: Not applicable.

Conflicts of Interest: The authors declare no conflict of interest.

References

1. Basu, R.; Billa, L.R.; Rao, J.M.; Letizia, R.; Paoloni, C. Design of D-band Double Corrugated Waveguide TWT for Wireless Communications. In Proceedings of the International Vacuum Electronics Conference (IVEC), Busan, Korea, 28 April–1 May 2019; pp. 1–2. [CrossRef]
2. Armstrong, C.M. The truth about terahertz. *IEEE Spectr.* **2012**, *50*, 36–41. [CrossRef]
3. Joye, C.D.; Kimura, T.; Hyttinen, M.; Levush, B.; Cook, A.M.; Calame, J.P.; Abe, D.K.; Vlasov, A.N.; Chernyavskiy, I.A.; Nguyen, K.T.; et al. Design of slow wave structure for G-band TWT for High Data Rate Links. *IEEE Trans. Electron. Devices* **2014**, *61*, 1672–1678. [CrossRef]
4. Yoshida, M.; Kobayashi, J.; Fujishita, Y.; Masuda, N.; Sekine, N.; Kanno, A.; Yamamoto, N.; Kasamatsu, A.; Hosako, I. Development activity of terahertz amplifiers with FWG-TWTs. In Proceedings of the 2016 IEEE International Vacuum Electronics Conference (IVEC), Monterey, CA, USA, 19–21 April 2016; pp. 1–2. [CrossRef]
5. Shi, N.; Zhang, C.; Wang, S.; Tian, H.; Shao, W.; Wang, Z.; Tang, T.; Duan, Z.; Lu, Z.; Gong, H.; et al. A Novel Scheme for Gain and Power Enhancement of THz TWTs by Extended Interaction Cavities. *IEEE Trans. Electron. Devices* **2020**, *67*, 1–6. [CrossRef]
6. Basten, M.A.; Tucek, J.C.; Gallagher, D.A.; Kreischer, K.E. 233 GHz high power amplifier development at northrop Grumman. In Proceedings of the 2016 IEEE International Vacuum Electronics Conference (IVEC), Monterey, CA, USA, 19–21 April 2016; pp. 1–2. [CrossRef]
7. Bian, B.X.; Pan, P.; Tang, Y.; Lu, Q.; Li, Y.; Zhang, L.; Wu, X.; Cai, J.; Feng, J. Demonstration of a Pulsed G-Band 50-W Traveling Wave Tube. *IEEE Trans. Electron. Devices* **2021**, *42*, 248–251. [CrossRef]
8. Baig, A.; Gamzina, D.; Kimura, T.; Atkinson, J.; Domier, C.; Popovic, B.; Himes, L.; Barchfeld, R.; Field, M.; Luhmann, N.C. Performance of a nano-CNC machined 220-GHz traveling wave tube amplifier. *IEEE Trans. Electron. Devices* **2017**, *64*, 2390–2397. [CrossRef]
9. Lei, W.; Hu, P.; Huang, Y.; Jiang, Y.; Song, R.; Chen, H. A G-band wideband CW folded waveguide TWT. In Proceedings of the 2019 International Vacuum Electronics Conference (IVEC), Busan, Korea, 28 April–1 May 2019; pp. 1–2. [CrossRef]
10. Lei, W.; Jiang, Y.; Song, R.; Hu, P.; Chen, H.; Zhang, L.; Ma, G. Progress of G-band CW transformed folded waveguide TWT. In Proceedings of the 2021 22nd International Vacuum Electronics Conference (IVEC), Rotterdam, The Netherlands, 27–30 April 2021; pp. 1–2. [CrossRef]
11. Pan, P.; Tang, Y.; Bian, X.; Zhang, L.; Lu, Q.; Li, Y.; Feng, Y.; Feng, J. A G-band traveling wave tube with 20 W continuous wave output power. *IEEE Electron. Device Lett.* **2020**, *41*, 1833–1836. [CrossRef]
12. Liu, S.; Cai, J.; Feng, J.; Wu, X. Characteristics Study of 3π Stopbands of Folded Waveguide Slow-Wave Structures in V-Band Traveling-Wave Tubes. *IEEE Trans. Electron. Devices* **2016**, *63*, 1294–1298. [CrossRef]

Article

A 66–76 GHz Wide Dynamic Range GaAs Transceiver for Channel Emulator Application

Peigen Zhou *, Chen Wang, Jin Sun, Zhe Chen, Jixin Chen and Wei Hong

State Key Laboratory of Millimeter Waves, School of Information Science and Engineering, Southeast University, Nanjing 210096, China; chenwang@seu.edu.cn (C.W.); 220200844@seu.edu.cn (J.S.); zhechen@seu.edu.cn (Z.C.); jxchen@seu.edu.cn (J.C.); weihong@seu.edu.cn (W.H.)
* Correspondence: pgzhouseu@seu.edu.cn

Abstract: In this study, we developed a single-channel channel emulator module with an operating frequency covering 66–67 GHz, including a 66–76 GHz wide dynamic range monolithic integrated circuit designed based on 0.1 μm pHEMT GaAs process, a printed circuit board (PCB) power supply bias network, and low-loss ridge microstrip line to WR12 (60–90 GHz) waveguide transition structure. Benefiting from the on-chip multistage band-pass filter integrated at the local oscillator (LO) and radio frequency (RF) ends, the module's spurious components at the RF port were greatly suppressed, making the module's output power dynamic range over 50 dB. Due to the frequency-selective filter integrated in the LO chain, each clutter suppression in the LO chain exceeds 40 dBc. Up and down conversion loss of the module is better than 14 dB over the 66–67 GHz band, the measured IF input P1 dB is better than 10 dBm, and the module consumes 129 mA from a 5 V low dropout supply. A low-loss ridged waveguide ladder transition was designed (less than 0.4 dB) so that the output interface of the module is a WR12 waveguide interface, which is convenient for direct connection with an instrument with E-band (60–90 GHz) waveguide interface.

Keywords: channel emulator; E-band; GaAs; ridged waveguide ladder transition; wide dynamic range

1. Introduction

6G is a new generation of mobile communication system developed for the needs of future mobile communication. According to the law of mobile communication development, 6G will have ultra-high spectrum utilization and energy efficiency, and will be an order of magnitude or higher more efficient than 5G in terms of transmission rate and spectrum utilization. Its wireless coverage performance, transmission delay, system security, and user experience will also be significantly improved [1]. Traditional mobile communications mainly use the frequency band below 6 GHz, which has become increasingly saturated. With the huge demand of 6G for system capacity, in some 6G applications, spectrum resources of several GHz may be required to meet specific transmission requirements. Due to the abundance of spectrum resources in the millimeter-wave (mm-wave) band, the World Radio Conference (WRC-19) has approved multiple mm-wave spectra for future mobile communications research and development, including the 66–67 GHz frequency range [2].

Different from frequency bands below 6G, the mm-wave band has poor penetration ability and large path loss. In order to effectively increase the propagation distance of the link, an effective solution is to adopt a large-scale multiple-input multiple-output (MIMO) system architecture [3]. In a MIMO system architecture, the number of antennas is greatly increased, and the systematic conductive testing becomes more impractical due to the influence of the long calibration time [4]. Moreover, the cost of testing is expensive, and the complexity of the test system increases exponentially. Therefore, multi-probe over-the-air (OTA) testing in millimeter-wave shielded anechoic chambers is becoming the mainstream testing solution [5–7].

As a vital part of the mm-wave OTA test system, a wireless channel emulator can accurately simulate and measure the wireless environment's degradation of system performance, including the free space path loss, shadowing, and multi-path fading of the transmitted signals from the antenna ports [4]. Transceivers in the channel emulator send/receive up/down mm-wave signals to/from OTA probe antennas. The channel emulator is an important high-end general-purpose instrument for verifying the performance of system equipment and terminal equipment in various complex channel environments. However, due to high demands on dynamic range, compactness, cost, and power consumption, the mm-wave channel emulators are rarely seen, especially when operating at frequencies above 50 GHz. At present, the known channel emulator with the highest operating frequency is PROPSIM released by Keysight, which can support up to 43.5 GHz [8].

This paper presents a 66–67 GHz transceiver monolithic microwave integrated circuit (MMIC) in waveguide module for massive MIMO channel emulator application. The proposed transceiver is integrated by cascading a tripler chain for LO drive, a mixer, and a band-pass filter using a 0.1 μm pHEMT GaAs process.

A high dynamic output power range, up to 50 dB over 66-to-76 GHz, is achieved by carefully dealing with all unwanted harmonic signals employing highly selective band-pass and high-pass filters in the transceiver. Total power consumption of the chip is 645 mW with the supply voltage of 5 V. A low-loss ridged waveguide ladder transition was designed so that the output interface of the module is a WR12 waveguide interface, which is convenient for direct connection with an instrument with E-band waveguide interface. To the best of our knowledge, the proposed integrated module is a competitive E-band transceiver system for radio channel emulation application.

2. System Architecture

Figure 1 shows the system block diagram of the channel emulator and its connection to an RF system. The whole N-channel system includes three parts: baseband, LO signal source, IF chain, and RF front-end. Among them, the research on baseband (BB), Lo signal source, and IF TX/RX is relatively mature. However, there are few research reports on RF channel emulators. An E-band specific RF front-end chip for channel emulator applications is currently scarce in the market. Therefore, in this paper we develop a channel simulator for evaluating E-band channel characteristics for future E-band communication application scenarios.

The red dotted box in Figure 1 shows the system block diagram of the 66–67 GHz channel emulator module designed in this paper. The transceiver chip is realized by cascading a frequency tripler, the first LO band-pass filter (BPF), a LO driver amplifier, the LO high-pass filter (HPF), a mixer, and the second RF BPF [4]. The LO chain integrates a tripler instead of a doubler or higher order frequency multiplier in order to make a compromise between test convenience, conversion gain, and power consumption. Additionally, the tripler was chosen to have a trade-off between the cost of the LO signal source and power consumption. The IF frequency is selected around 27 GHz in the 5G mm-wave hotspot frequency band to facilitate the measurement of the channel emulator. In order to facilitate the connection with the instrument and the TX module, the IF and LO are coaxial interfaces, and the LO and IF ports of the transceiver are connected to the module via microstrip gold wire bonding. The RF output port of the module is a WR12 waveguide interface, as shown in Figure 1.

The mixer in the system architecture is a passive star mixer, so the channel emulator can be used for TX testing or RX testing. When applied to TX test scenarios, the RF signal output power budget is between −50 dBm and 0 dBm over 66-to-76 GHz. When an RX is tested, the input power range of the RF signal is −45~0 dBm. In addition, for the channel emulator, the transmit output power and receive noise figure are not key indicators, so the RF power amplifier and low noise amplifier are not integrated in the system [7,8]. To the best of our knowledge, extensive research has been carried out on mm-wave transceivers [9–12], while studies have rarely been published concerning E-band radio channel emulator application.

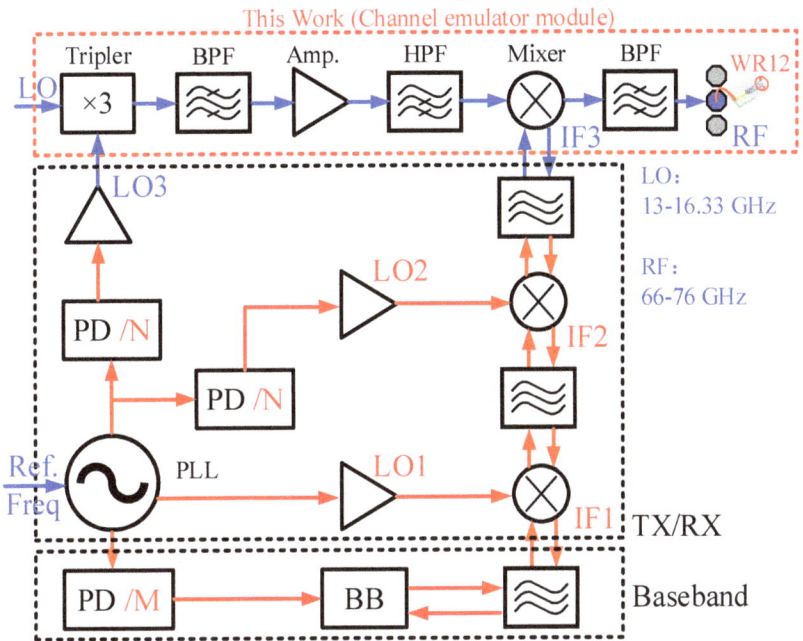

Figure 1. System block diagram of the channel emulator system.

3. Circuit Design Methodology

In this section, we will present the circuit design methodology of the 66–67 GHz channel emulator, including the consideration of each block, and then present the simulated and measured results of key building blocks.

3.1. Frequency Tripler

As shown in Figure 1, the LO chain integrates a frequency tripler, a BPF, an LO driving power amplifier, and an HPF. The schematic of the frequency tripler and the succeeding frequency selective BPF is illustrated in Figure 2. The tripler core is composed of anti-parallel diode pairs (APDP) [13,14], and the diode is implemented by connecting the drain and source of pHEMT transistors as the cathode with the gate as the anode. Due to the passive structure, sufficient input power is required for the tripler to generate odd-order harmonics while suppressing even-order spurs [15]. Two 4-finger 10 μm pHEMTs are employed as the APDP in this design in consideration of a trade-off between output power and bandwidth. The input matching network consists of a capacitor connected in parallel to ground and a microstrip line connected in series. The output matching network consists of two capacitors connected in series and an inductor connected in parallel to ground in a T-shaped configuration. In order to effectively improve the unwanted harmonics suppression of the tripler and reduce the frequency conversion loss, a transmission line TL as depicted in Figure 2 is adopted to reflect idle frequency signals to the APDP core. A seventh-order BPF was connected after the frequency multiplier to further improve the harmonic suppression characteristics of the LO chain.

The simulation results of each harmonic output power of the tripler plus the cascaded BPF when the input power is 16 dBm are shown in Figure 3. The simulated input and output return loss are better than −10 dB over 41–51 GHz, and the simulated output power of the third harmonic signal is around 0 dBm. Benefiting from the BPF, all unwanted harmonics can be suppressed significantly within the interested frequency bands. It can be deduced from Figure 3 that the fundamental signal and 2nd harmonic rejection are over 30 dB compared with the third harmonic signal, and the 4th and 5th harmonic suppressions are better than 35 dB.

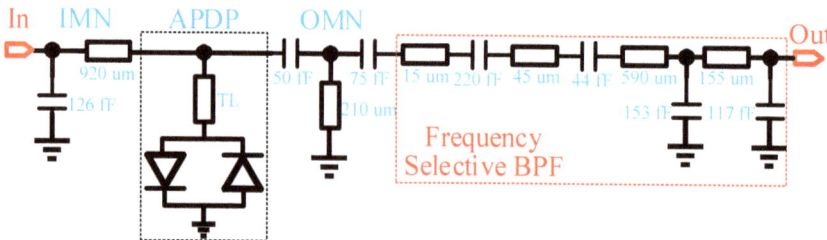

Figure 2. Schematic of the frequency tripler and its cascaded frequency selective BPF.

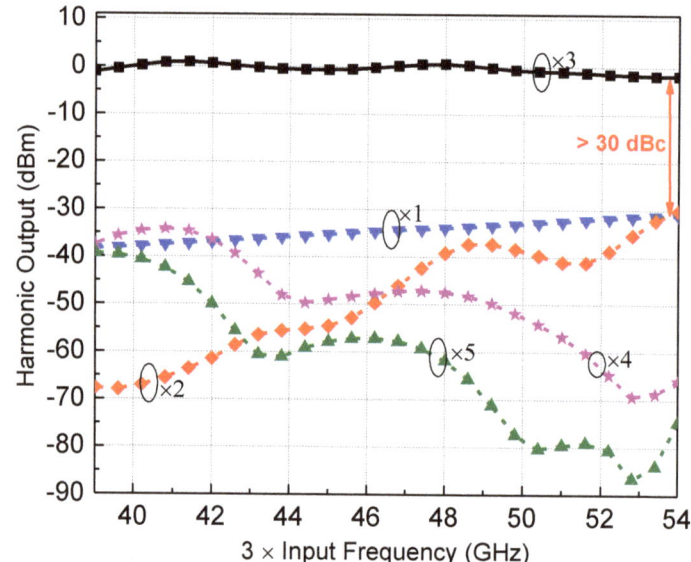

Figure 3. Simulated output power of the tripler with BPF.

3.2. Power Amplifier

The schematic of the LO driving power amplifier and the HPF is shown in Figure 4. The power amplifier adopts a three-stage common-source structure to obtain sufficient power gain at 41–51 GHz while ensuring high power-added efficiency (PAE) [16]. The first two driver stages use a 4 × 25 μm pHEMT transistor to obtain sufficient gain, and the final power stage uses a larger 4 × 50 μm pHEMT transistor to obtain a sufficiently high output power. Source degeneration inductors are connected to the source of the pHEMT transistors to increase the stability of the power amplifier. The input matching network of the power amplifier is co-designed with the previous frequency selective BPF. The output matching network is co-simulated with the following HPF in full wave electromagnetic simulation. All transistors are biased with a shunt by-pass capacitor and a series resistor close to the gate, and the dc power (Vdd) is feeded through an LC network to the drain.

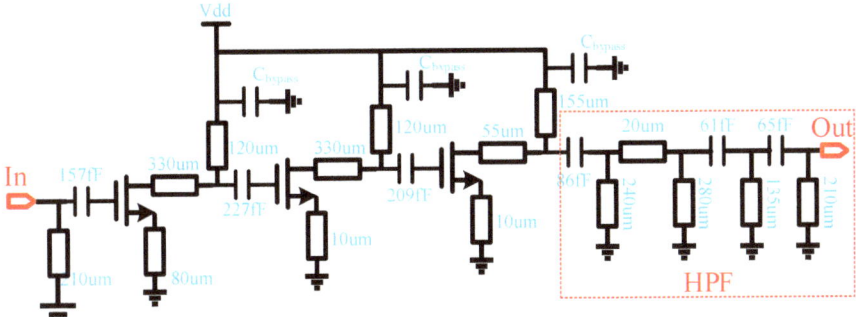

Figure 4. Schematic of the power amplifier with HPF.

For the power amplifier used in the LO chain, a key indicator is the out-of-band suppression. For the channel emulator, the requirements for the suppression of each harmonic in the LO chain are higher, because this out-of-band clutter will degrade the dynamic operating power range of the channel emulator. Therefore, the fifth-order HPF is employed after the power amplifier. As depicted in Figure 5, an additional 25 dBc forward fundamental signal rejection can be obtained with the HPF. The 2nd harmonic is suppressed by 10–25 dBc, while the 4th and 5th harmonics are amplified with low gain by tuning matching networks to make the out-of-band gain drop slope as steep as possible. The simulated input and output return loss of the power amplifier is better than −10 dB in the 41–51 GHz frequency band, and the small-signal gain is around 20 dB. Saturated output power varies from 17 dBm to 18.8 dBm including the insertion loss of the HPF. The circuit draws a total current of 102 mA at 3.3 V power supply with −0.4 V gate bias.

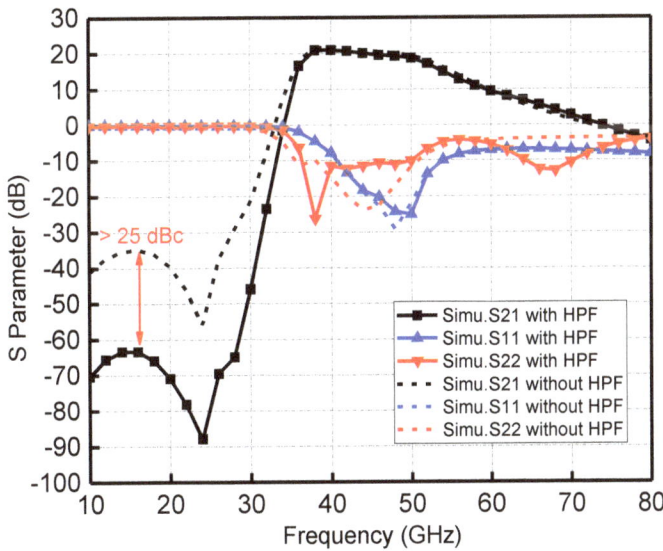

Figure 5. Simulated S-parameters of the LO driver power amplifier with or without HPF.

3.3. Star Mixer

The passive star mixer has been widely used in mm-wave on-chip systems since being proposed by Basu and Maas [17]. Its highly symmetrical star topology can offer better port-to-port isolation and high spurious rejection [18,19]. Different from the traditional star mixer, the traditional straight symmetrical Marchand Balun is modified to S type with two bends, which can reduce the chip size while ensuring the performance of the mixer, as shown in Figure 6. To reduce the coupling between transmission lines, the decoupling ground wall consisting of metal-vias array is inserted between four S type balun and the IF port. Additionally, double side coupling lines of the Marchand Balun are grounded separately in the end, rather than joint together by the bottom metal layer for better spurious rejection. Four diodes measuring 1×15 μm are arranged in a symmetrical layout for broad IF bandwidth. In addition, the LO input matching network is co-optimized with a pre-stage HPF output network. Furthermore, an extra BPF with the same topology as the one after the tripler is integrated after the star mixer. Finally, the micrograph of the complete channel emulator is shown in Figure 7, with a chip size of 2.7 by 0.9 mm^2.

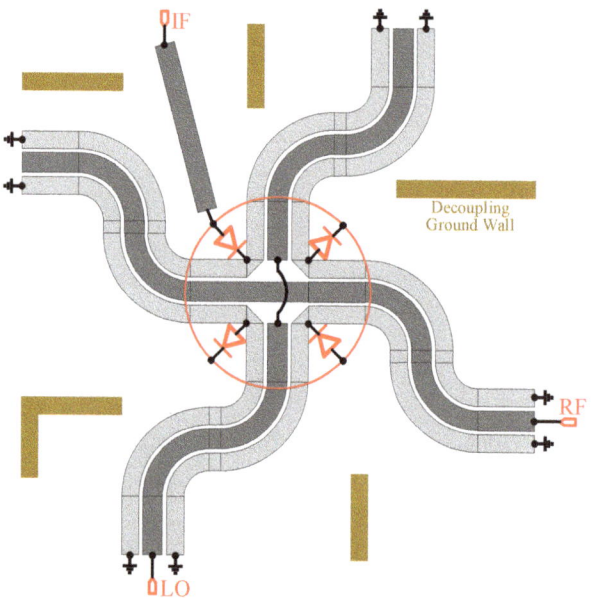

Figure 6. Schematic of the modified star mixer.

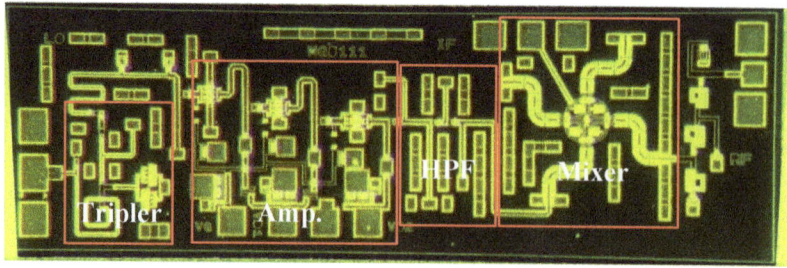

Figure 7. Chip micrograph of the channel emulator.

Simulation results of the mixer with the following BPF show that both the input and output return loss is better than -10 dB, and the conversion loss of the mixer is about 6.5 dB from 66 to 76 GHz. Furthermore, the insertion loss of the BPF after the mixer is between -1.4 and -2 dB within the RF operating bandwidth.

4. Packaging Methodology

As an important part of the OTA test system, the channel emulator module is directly connected to the RF transceiver of the communication system to simulate path loss, multipath fading, and shadow fading in the wireless environment. Therefore, the input and output interface of the channel emulator should have better robustness so that it can be directly connected with the transceiver or instrument. In mm-wave low-frequency bands (below 30 GHz), interfaces such as instrument or RF transceiver outputs typically use coaxial connectors [20]. When the frequency is higher than 50 GHz, the instrument interface is usually designed as a waveguide interface for more favorable stability and cost.

This channel emulator is aimed at 66–67 GHz wireless communication system test applications, and the RF frequency range is 66–67 GHz. The IF port is compatible with the existing commercial 5G mm-wave band, the IF frequency is 27 GHz, and the IF power coverage range is −40~10 dBm. The LO input frequency range is 13~16.33 GHz. Therefore, the LO and the IF port use a coaxial connection scheme, and the RF port uses a WR12 waveguide interface. The frequency of the IF and LO ports is lower, and the influence of the gold bonding wire is small. In the module design, the RF and IF signal interfaces on the chip are directly bonded to the PCB through gold wires, and are connected to the coaxial connector through the 50 ohm characteristic impedance microstrip line on the PCB.

The working frequency of the RF port is relatively high. In order to transfer the RF port to the WR12 waveguide interface, firstly, the RF GSG PAD-to-50ohm microstrip line (as shown in Figure 8) was designed on the TLY-5 sheet with a dielectric constant of 2.2 and a thickness of 0.254 mm using bonding wires. The distance from the microstrip line to the edge of the RF GSG PAD, the height of the gold bonding wire, and the structure of the microstrip line were all optimized by electromagnetic simulation [21]. Then, a low-loss microstrip to WR12 ridged waveguide ladder transition (RWLT) structure (as shown in Figure 9) was designed using aluminum metal [22,23]. The transition structure includes a 4-stage stepped impedance transformation. By optimizing the length and height of each stepped transformation unit, a lower transition loss from the microstrip to WR12 can be obtained, and the connection loss in the 66–67 GHz frequency band is lower than 0.36 dB.

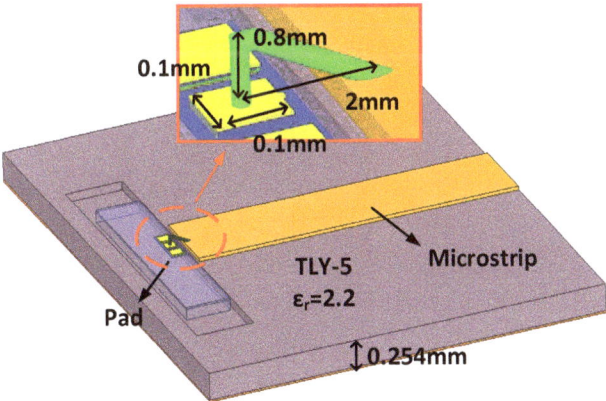

Figure 8. 3D model of the RF GSG PAD to microstrip transition.

Figure 10 shows the S-parameters of the transition structure between the simulated RF GSG PAD and the microstrip line and the microstrip line to the WR12 waveguide port in the entire E-band. As illustrated in Figure 10, in the channel emulator operating frequency band 66–67 GHz, the return losses (S11, S22) of the above connection structures are all better than −10 dB, and the total insertion loss between the RF GSG PAD and the WR12 waveguide port is equal to less than 2 dB.

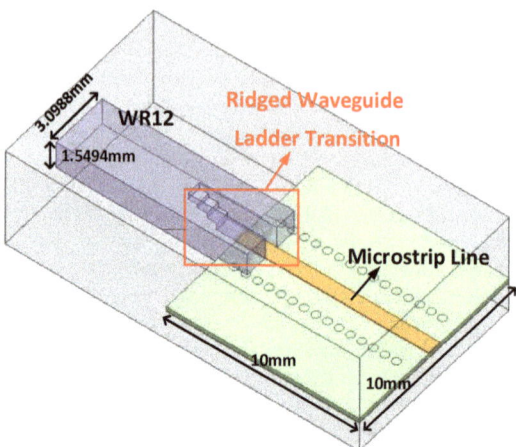

Figure 9. Ridge waveguide ladder transition from WR12 waveguide interface to 50-ohm microstrip line.

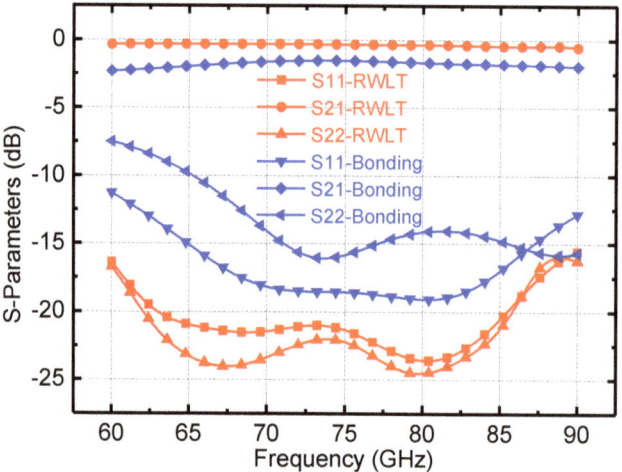

Figure 10. The simulated connection loss of the RF GSG PAD to the microstrip line and the microstrip line to the WR12 waveguide port.

The photograph of the opened split-block and assembled module including the RF transceiver chip, RF microstrip-to-waveguide transition, and the DC biasing network is shown in Figure 11. The PCB was sintered on the aluminum structure to obtain a tighter and better grounding effect. The RF transceiver chip was installed in a groove dug in the middle of the PCB. The depth of the groove was slightly higher than the height of the chip to ensure that the top surface of the chip (after the conductive adhesive is pasted) was the same height as the PCB surface. In the design of the DC bias network, we first placed some decoupling chip capacitors next to the chip to obtain better DC bias characteristics. The DC voltage of the amplifier's drain, gate, etc. is provided through the LDO DC biasing network, and the entire module has only a 5V DC input voltage. On the side of the module, two 2.92-mm coaxial connectors were used as LO and IF ports, and the RF port is a WR12 waveguide interface.

Figure 11. Photo of the 66–67 GHz channel emulator module integrating the RF transceiver chip, the microstrip-to-waveguide transition structure, and the DC biasing network.

5. Measurement Results

The measurements were performed using two steps. First, the RF transceiver chip was measured with on-chip characterization to obtain accurate conversion gain, dynamic range, and other performances; then, the packaged module was tested. In addition, since the harmonic suppression performance of the LO link is important for the channel emulator, the LO chain was separately processed and the output power of each harmonic was tested; the measured results are illustrated in Figure 12. In the working frequency range of $3 \times LO_{In}$ (39–49 GHz), the output power of the third harmonic of the LO chain exceeds 15 dBm, and the unwanted harmonics suppression of each order exceeds 40 dBc.

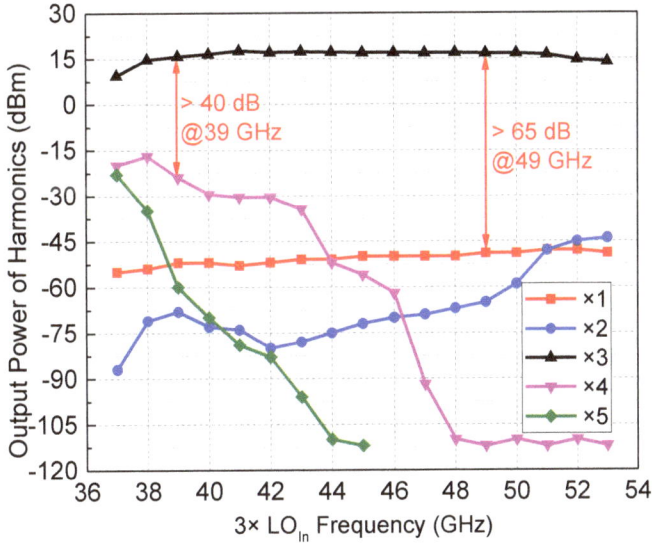

Figure 12. The measured output power of each harmonic of the LO chain.

The on-chip measurement setup of an up-conversion configuration is shown in Figure 13. In this measurement, the chip is implemented as a transmitter. LO and IF signals are pumped from two signal source Keysight E8257Ds through co-axial cables and probes. The RF signal is measured by a waveguide probe and a spectrum analyzer. As our corresponding band was E-band, the measurements were conducted by two steps and limited by our testing equipment. One is the V-band (50–75 GHz) test and the other is the W-band (75–110 GHz) measurement. For the two setups, different waveguide probes and two kinds of harmonic mixers were applied. To obtain the accurate output power of the chip, power meter was employed. The down-conversion test setup was similar to that in Figure 13, expect that the RF output was changed to input, and the IF input was changed to output.

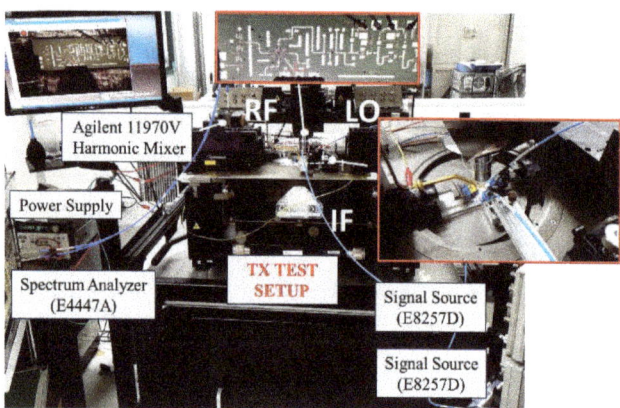

Figure 13. On-chip measurement setup.

Figure 14 shows the measured results of the conversion gain of the RF transceiver chip and the channel emulator module when the LO signal power is 16 dBm and the IF signal frequency is 27 GHz. The measured results show that the conversion gain of the RF chip is between −8 and −12 dB in the 66–67 GHz frequency band. The conversion gain corresponding to the transmit mode and the receive mode are in agreement. The conversion loss measured in the 66–67 GHz band is large, mainly because the output power of the LO chain is low, which is not enough to drive the star mixer. The conversion loss of the channel emulator module integrated with the RF waveguide interface and the LO and IF coaxial interfaces is about 2 dB higher than the on-chip measured results. In addition to the transition between the RF GSG PAD and the WR12 waveguide port, this part of the loss also includes the loss of a section of the IF signal transmission line on the PCB.

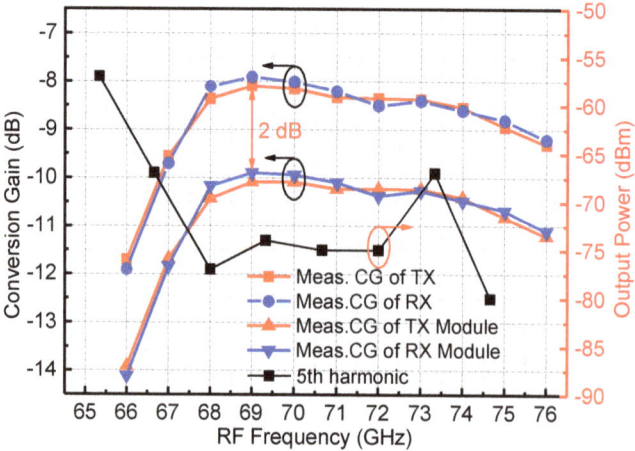

Figure 14. Measured conversion gain of the RF channel emulator and the module.

Additionally, the in-band harmonic signal has a great influence on the sensitivity of the channel emulator. In this design, the frequency of the IF signal is 27 GHz, the frequency of the LO signal is 13–16.33 GHz, and the frequency of the RF signal is 66–67 GHz. The 5th harmonic of the LO signal falls within the RF band. Therefore, we measured the power of the 5th harmonic of the LO input signal in the RF band, and the results are shown in Figure 14. It can be deduced that within the RF bandwidth, the 5th harmonic leakage amplitude of the LO is less than −63 dBm.

Figure 15 shows the measured results of the RF output power of the channel simulator module varying with the IF input power. The measured input P1 dB of the module is about 10 dBm. In the 66–67 GHz band, the RF output 1 dB compression point is between −2 and −4.5 dBm. For linear channel simulator transceiver application scenarios, considering a certain spurious suppression margin, the dynamic operating power range of the module can reach more than 50 dB. In the down-conversion test, the receiving dynamic range can reach more than 54 dB under the same signal quality.

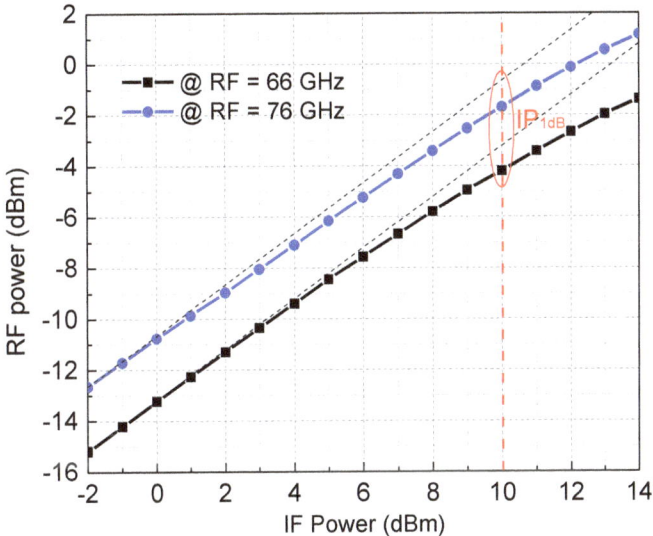

Figure 15. The measured RF output power of the channel emulator varies with the IF input power.

6. Conclusions

This paper presents a 66–67 GHz channel emulator module in a waveguide package. The module includes a 66–67 GHz transceiver chip processed by a 0.1 μm pHEMT GaAs process, a DC bias network designed on a PCB, and a low-loss ridge microstrip line to a WR12 waveguide transition structure. In each circuit block of the transceiver, the various clutter signals are closely monitored and studied. By integrating multiple frequency-selective filters and a high-isolation mixer in the link, the transceiver achieves good spurious rejection performance. In addition, benefiting from the designed low-loss stepped microstrip-to-waveguide transition structure, the RF output of the module is a WR12 waveguide interface. This makes it easy to interface with commercial E-band instruments to evaluate the performance of the channel emulator module. Measured results show that the channel emulator module can achieve a dynamic operating power range of more than 50 dB in the 66 to 76 GHz frequency band. Due to better dynamic range performance and higher integration, the transceiver chip and module can meet the application requirements of E-band channel simulators.

Author Contributions: Conceptualization, P.Z., C.W. and W.H.; methodology, P.Z., C.W., J.S., Z.C. and J.C.; validation, P.Z., C.W. and J.S.; formal analysis, P.Z., C.W. and J.S.; data curation, P.Z., C.W. and J.S.; writing—original draft preparation, P.Z.; writing—review and editing, P.Z.; supervision, J.C. and W.H. All authors have read and agreed to the published version of the manuscript.

Funding: This work was supported in part by the Natural Science Foundation of Jiangsu Province under Grant BK20210206, in part by the National Natural Science Foundation of China under Grant 62101117 and 62188102, and in part by the Project funded by China Postdoctoral Science Foundation under Grant 277393.

Data Availability Statement: Not applicable.

Conflicts of Interest: The authors declare no conflict of interest.

References

1. Wang, C.-X.; Huang, J.; Wang, H.; Gao, X.; You, X.; Hao, Y. 6G Wireless Channel Measurements and Models: Trends and Challenges. *IEEE Veh. Technol. Mag.* **2020**, *15*, 22–32. [CrossRef]
2. ITU. Provisional Final Acts WRC-19. White Paper. Available online: http://handle.itu.int/11.1002/pub/813801e6-en (accessed on 1 March 2022).
3. Yang, B.; Yu, Z.; Lan, J.; Zhang, R.; Zhou, J.; Hong, W. Digital Beamforming-Based Massive MIMO Transceiver for 5G Millimeter-Wave Communications. *IEEE Trans. Microw. Theory Tech.* **2018**, *66*, 3403–3418. [CrossRef]
4. Wang, C.; Hou, D.; Zheng, S.; Chen, J.; Zhang, N.; Jiang, Z.; Hong, W. E-band transceiver monolithic microwave integrated circuit in a waveguide package for millimeter-wave radio channel emulation applications. *Sci. China Inf. Sci.* **2022**, *65*, 129404. [CrossRef]
5. Zhang, P.; Yang, X.; Chen, J.; Huang, Y. A survey of testing for 5G: Solutions, opportunities, and challenges. *China Commun.* **2019**, *16*, 69–85.
6. Scannavini, A.; Foged, L.J.; Gross, N.; Noren, P. OTA measurement of wireless devices with single and multiple antennas in anechoic chamber. In Proceedings of the 8th European Conference on Antennas and Propagation (EuCAP), The Hague, The Netherlands, 6–11 April 2014; pp. 3495–3499.
7. Mehmood, R.; Wallace, J.W.; Jensen, M.A. Reconfigurable OTA chamber for MIMO wireless device testing. In Proceedings of the 2016 10th European Conference on Antennas and Propagation (EuCAP), Davos, Switzerland, 10–15 April 2016; pp. 1–4.
8. Keysight. PROPSIM F64 Radio Channel Emulator F8800A. Datasheet. Available online: https://www.keysight.com/us/en/assets/7018-06665/data-sheets/5992-4078.pdf (accessed on 1 March 2022).
9. Levinger, R.; Yishay, R.B.; Katz, O.; Sheinman, B.; Mazor, N.; Carmon, R.; Elad, D. High-Performance E-Band Transceiver Chipset for Point-to-Point Communication in SiGe BiCMOS Technology. *IEEE Trans. Microw. Theory Tech.* **2016**, *64*, 1078–1087. [CrossRef]
10. Dyadyuk, V.; Shen, M.; Stokes, L. An E-band transceiver with 5 GHz IF bandwidth. In Proceedings of the 2014 1st Australian Microwave Symposium (AMS), Melbourne, Australia, 26–27 June 2014; pp. 43–44.
11. Tokgoz, K.K.; Maki, S.; Pang, J.; Nagashima, N.; Abdo, I.; Kawai, S.; Fujimura, T.; Kawano, Y.; Suzuki, T.; Iwai, T.; et al. A 120 Gb/s 16QAM CMOS millimeter-wave wireless transceiver. In Proceedings of the 2018 IEEE International Solid-State Circuits Conference—(ISSCC), San Francisco, CA, USA, 11–15 February 2018; pp. 168–170.
12. Kim, H.T.; Park, B.S.; Song, S.S.; Moon, T.S.; Kim, S.H.; Kim, J.M.; Chang, J.Y.; Ho, Y.C. A 28-GHz CMOS Direct Conversion Transceiver with Packaged 2 × 4 Antenna Array for 5G Cellular System. *IEEE J. Solid-State Circuits* **2018**, *53*, 1245–1259. [CrossRef]
13. Shim, D.; Choi, W.; Lee, J.W. Self-Dynamic and Static Biasing for Output Power and Efficiency Enhancement of Complementary Antiparallel Diode Pair Frequency Tripler. *IEEE Microw. Wirel. Compon. Lett.* **2017**, *27*, 1110–1112. [CrossRef]
14. Hung, C.; Chiong, C.; Chen, P.; Tsai, Y.; Tsai, Z.; Wang, H. A 72–114 GHz fully integrated frequency multiplier chain for astronomical applications in 0.15-μm mHEMT process. In Proceedings of the 40th European Microwave Conference, Paris, France, 28–30 September 2010; pp. 81–84.
15. Zhou, P.; Chen, J.; Yan, P.; Yu, J.; Hou, D.; Gao, H.; Hong, W. An E-Band SiGe High Efficiency, High Harmonic Suppression Amplifier Multiplier Chain with Wide Temperature Operating Range. *IEEE Trans. Circuits Syst. I Regul. Pap.* **2022**, *69*, 1041–1050. [CrossRef]
16. Hou, D.; Chen, J.; Yan, P.; Hong, W. A 270 GHz × 9 Multiplier Chain MMIC with On-Chip Dielectric-Resonator Antenna. *IEEE Trans. Terahertz Sci. Technol.* **2018**, *8*, 224–230. [CrossRef]
17. Basu, S.; Maas, S.A. Design and performance of a planar star mixer. *IEEE Trans. Microw. Theory Tech.* **1993**, *41*, 2028–2030. [CrossRef]
18. Kyung-Whan, Y.; Du-Hyun, K. A novel 60-GHz monolithic star mixer using gate-drain-connected pHEMT diodes. *IEEE Trans. Microw. Theory Tech.* **2005**, *53*, 2435–2440. [CrossRef]
19. Kuo, C.; Kuo, C.; Kuo, C.; Maas, S.A.; Wang, H. Novel Miniature and Broadband Millimeter-Wave Monolithic Star Mixers. *IEEE Trans. Microw. Theory Tech.* **2008**, *56*, 793–802.
20. Zhou, P.; Chen, J.; Yan, P.; Yu, J.; Li, H.; Hou, D.; Gao, H.; Hong, W. A 150-GHz Transmitter with 12-dBm Peak Output Power Using 130-nm SiGe: C BiCMOS Process. *IEEE Trans. Microw. Theory Tech.* **2020**, *68*, 3056–3067. [CrossRef]

21. Zhou, P.; Chen, J.; Yan, P.; Hou, D.; Hong, W. A low power, high sensitivity SiGe HBT static frequency divider up to 90 GHz for millimeter-wave application. *China Commun.* **2019**, *16*, 85–94.
22. Herraiz, D.; Esteban, H.; Martínez, J.A.; Belenguer, A.; Boria, V. Microstrip to Ridge Empty Substrate-Integrated Waveguide Transition for Broadband Microwave Applications. *IEEE Microw. Wirel. Compon. Lett.* **2020**, *30*, 257–260. [CrossRef]
23. Voineau, F.; Ghiotto, A.; Kerhervé, E.; Sié, M.; Martineau, B. Broadband 55–95 GHz microstrip to waveguide transition based on a dielectric tip and a tapered double-ridged waveguide section. In Proceedings of the 2017 IEEE MTT-S International Microwave Symposium (IMS), Honolulu, Hawaii, 4–9 June 2017; pp. 723–726.

Article

Compact Wideband Four-Port MIMO Antenna for Sub-6 GHz and Internet of Things Applications

Nathirulla Sheriff [1,*], Sharul Kamal [1], Hassan Tariq Chattha [2], Tan Kim Geok [3] and Bilal A. Khawaja [4]

1 Wireless Communication Center, School of Electrical Engineering, Faculty of Engineering, Universiti Teknologi Malaysia, Johor Bahru 81310, Malaysia
2 Advanced Cyclotron Systems Inc. (ACSI), Richmond, BC V6X 1X5, Canada
3 Faculty of Engineering and Technology, Multimedia University, Melaka 75450, Malaysia
4 Department of Electrical Engineering, Faculty of Engineering, Islamic University of Madinah, P.O. Box 170, Madinah 41411, Saudi Arabia
* Correspondence: nadhir_sh@live.com

Abstract: A compact four-port multi-input, multi-output (MIMO) antenna with good isolation is proposed for sub-6 GHz and Internet of Things (IoT) applications. Four similar L-shaped antennae are placed orthogonally at 7.6 mm distance from the corner of the FR4 substrate. The wideband characteristics and the required frequency band are achieved through the L-shaped structure and with proper placement of the slots on the substrate. To obtain good isolation between the ports, rectangular slots are etched in the bottom layer and are interconnected. The proposed antenna has total dimensions of 40 mm × 40 mm × 1.6 mm. The interconnected ground plane provides good isolation of less than −17 dB between the ports, and the impedance bandwidth obtained by the proposed four-port antenna is about 54% between the frequency range of 3.2 GHz to 5.6 GHz, thus providing a wideband antenna characteristic covering sub-6 GHz 5G bands (from 3.4 to 3.6 GHz and 4.8 to 5 GHz) and the WLAN band (5.2 GHz). The proposed design antenna is fabricated and tested. Good experimental results are achieved when compared with the simulation results. As the proposed design is compact and low profile, this antenna could be a suitable candidate for 5G and IoT devices.

Keywords: 5G; MIMO; IoT devices; Wi-Fi; four-ports; sub-6 GHz

1. Introduction

With a large number of users and the rapid development of wireless communication technologies, higher data rates and channel capacities are in great demand [1,2]. Multiple antennae integrating in the same portable device is seen as a hopeful solution, which could enhance communication network quality and channel capacity. Hence, multi-input, multi-output (MIMO) technology plays a key role in the 5G research hotspot. The European Commission (EC) announced that the band from 3.4 to 3.8 GHz was allocated for 5G, and similarly the Ministry of Industry and Information Technology of China has also considered 3.3–3.6 GHz and 4.8–5 GHz as the operation frequency bands of the 5G system [3]. Recently, many MIMO antenna designs for 5G sub-6 GHz were reported in the literature [4–12], but these antennae provide less bandwidth or higher mutual coupling. Contradictorily, the mutual coupling reduction and low envelope correlation coefficients (ECCs) between nearby antenna elements could increase the antenna size, and hence these factors play a key role in antenna design for portable devices. Hence, embedding multiple antennae inside the device in a limited space while maintaining good isolation becomes an antenna design challenge for portable devices.

Different techniques were presented in [13] to reduce the mutual coupling. In order to enhance the isolation, parasitic elements [14,15] are placed between radiating elements to create extra coupling paths. Defected ground structures [16] inhibit surface waves to reduce mutual coupling between the antenna elements by acting as band-stop filters. However, this technique decreases the total antenna efficiency. The etching of slots [17] disturbs the surface current distribution and the path length, which reduces the electromagnetic energy coupling between the ports. Neutralization lines [18] are employed for isolation enhancement by creating an extra coupling path suitable for narrow band decoupling. In [19,20] high isolation is achieved through the orthogonal polarization diversity technique using different excitation modes, while in [21] the multimode decoupling technique is employed to improve the isolation between the antenna elements. However, these designs work only for a single band or less bandwidth. In [7,9,22,23], an antenna is designed for multiple bands for sub-6 GHz applications. Good isolation is achieved using the slotted ground plane method in [24], and similarly in [25], the rectangular slot is etched in the ground plane to stop the flow of current. Moreover, the antenna designs presented were either complex in structure or larger in size and thus integration into a compact MIMO structure for portable devices could be challenging. Therefore, a unique antenna design with the features of extended bandwidth and good isolation suitable for sub-6 GHz and IoT applications needs to be investigated urgently.

In this paper, a compact four-port wideband MIMO antenna design is presented, with four antenna elements positioned near each other in a symmetric fashion with a common ground plane. Simple techniques of etching the slots are used in the top layer and the ground plane to attain the required impedance bandwidth and enhance the isolation between the ports. A peak gain lies between 2.4 to 4.9 dBi for the entire operational bandwidth and the average radiation efficiency obtained is 93%. ECC achieved is less than 0.05, which satisfies the IEEE standards [6] for MIMO antennae for portable devices. The impedance bandwidth obtained by the proposed four-port antenna is about 54%, which ranges from 3.2 GHz to 5.6 GHz, thus providing wideband antenna characteristics covering sub-6 GHz 5G bands (from 3.4 to 3.6 GHz and 4.8 to 5 GHz) and the WLAN band (5.2 GHz).

2. Antenna Design

A compact four-port MIMO antenna is designed and fabricated on a FR4 substrate with thickness (t) = 1.6 mm, loss tangent $(tan\delta)$ = 0.025, and dielectric constant (ε_r) = 4.4. Figure 1 illustrates the geometry of the proposed antenna, and the optimum parameters related to the proposed antenna design are listed in Table 1. A simple decoupling structure is implemented in the ground plane to obtain good isolation. CST Microwave Studio has been used for simulation purposes to design and analyze the antenna parameters. The total dimensions of the MIMO antenna are 40 mm × 40 mm (~0.59λ × 0.59λ at center frequency of 4.45 GHz). The design stages are demonstrated in the subsequent sections.

Table 1. Antenna dimensions of the proposed design (mm).

Parameter	Dimension (mm)
L	40
W	40
CSL	6.5
MSL	8
l_port	10
w_port	2.8
A	0.8
B	8
C	7
D	8.5
E	22
W_slot	10
L_slot	4

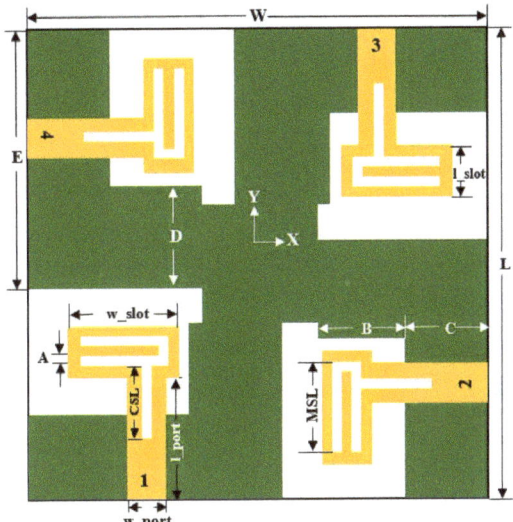

Figure 1. Structure and dimensions of proposed 4 × 4 MIMO antenna.

2.1. Single-Port Antenna

An initial configuration of the proposed MIMO system with a single-port antenna is shown in Figure 2C and its total dimensions is 20 mm × 20 mm × 1.6 mm. The design-evolution steps of the proposed antenna are shown in Figure 2. In Figure 3, the return-loss (S_{11}) results obtained in each evolution step while designing the proposed MIMO antenna are shown.

Initially, an antenna with less than half ground plane and an L-shaped antenna element is designed, as shown in Figure 2A. It can be observed that the impedance bandwidth of 26% is achieved from 4.6 to 6 GHz ($S_{11} < -10$ dB). In the next steps, the ground plane is modified (Figure 2B) and then further improved (Figure 2C) to radiate for the required frequency band. The S-parameter plot in Figure 3 shows that $S_{11} < -8$ dB is achieved for the complete required frequency range from 3.5–5.4 GHz with impedance bandwidth of 43%.

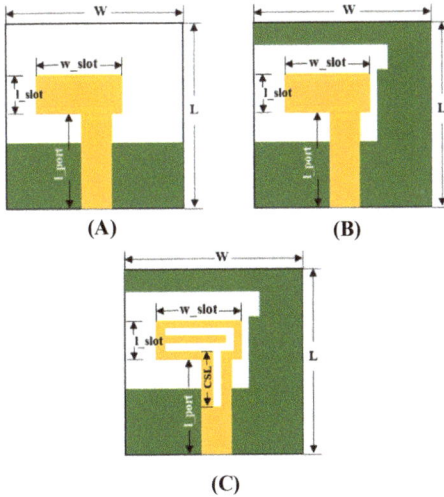

Figure 2. Design-evolution process of the single-element antenna. (**A**) Step–1, (**B**) Step–2, (**C**) Step–3.

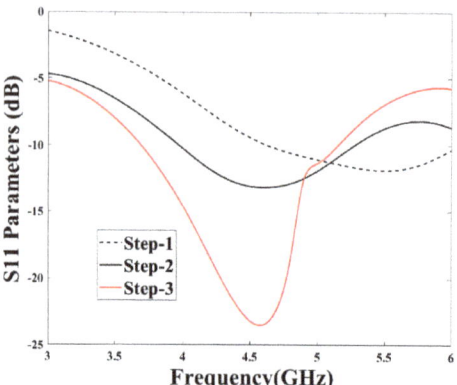

Figure 3. S–parameters for different configuration.

2.2. Four-Port MIMO Antenna

The four-port MIMO antenna is proposed from the single-port antenna design discussed in the preceding section. At the initial stage, four antennae are placed orthogonally to each other on the top layer of the FR4 substrate, as shown in Figure 4, and the total volume of the antenna is 40 mm × 40 mm × 1.6 mm. Each antenna element with its feeding port of width (w_port) = 3 mm is placed at a distance of 7.6 mm from the corner end of the substrate. The inter-element spacing between the two antenna elements is 12 mm. The dimensions are properly adjusted in such a way to achieve good bandwidth covering the required frequency range and to obtain good isolation.

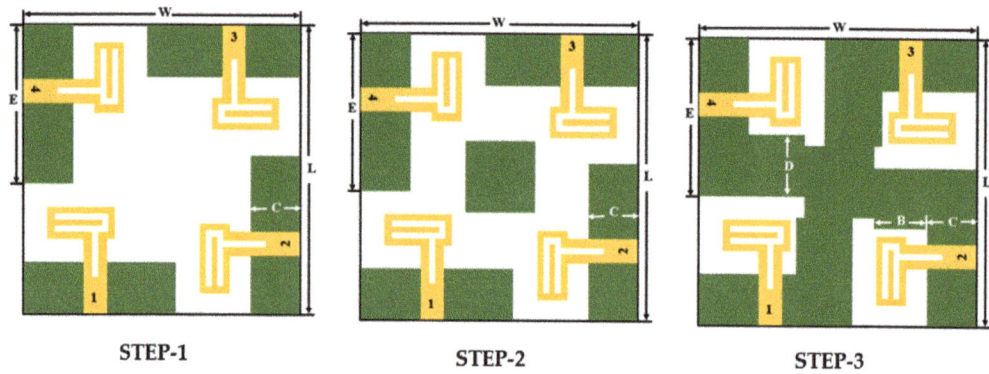

Figure 4. Design evolution of the ground plane structure for the proposed antenna.

The step-by-step configuration of the proposed antenna is shown in Figure 4. The simulated S-parameters plotted in Figure 5A clearly show that the return loss of −10 dB starts only from 4.5 GHz and from 4.2 GHz in steps 1 and 2, respectively. Similarly, both the step designs (STEP 1 and STEP 2) have high mutual coupling between the ports (Figure 5B). In order to achieve good isolation, the ground plane is modified as shown in Figure 4 (steps 2 and 3) by arranging a slot in the center of the antenna ground plane, connected to each other to form a common ground plane. It is also observed that, the currents almost penetrate between the nearby antenna elements in step 2 compared to current distribution in step 3. Good isolation and impedance bandwidth is achieved in the proposed design of step 3 (Figure 5B).

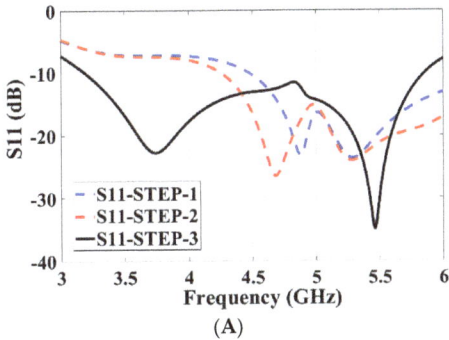
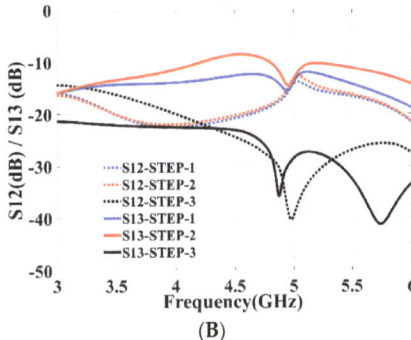

Figure 5. Frequency variation of S-parameters for different configuration: (A) S_{11} [dB], and (B) S_{13} [dB].

3. Results and Discussion

The fabricated four-port MIMO antenna is shown in Figure 6A (top view) and Figure 6B (bottom view). Using an Agilent PNA-X N5242A vector network analyzer (VNA), the S-parameters are measured. The radiation pattern is measured in an anechoic chamber by using a Nanjing Lopu Co. antenna measurement system. The measurement scenario of the proposed antenna is given in Figure 7 to show the measurement environment. The simulated and measured return loss for the designed antenna is represented in Figure 8A. The figure representation clearly shows that the simulated S_{11} results of all the four ports are the same due to its similar structure, and good impedance bandwidth of 57% is achieved between the frequency range of 3.2 GHz to 5.8 GHz.

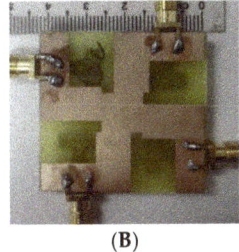

Figure 6. Fabricated prototype of the proposed antenna design: (A) top view and (B) bottom view.

On the other side, measured return-loss results of the proposed antenna show utmost similar results with good bandwidth of 54% covering the frequency range from 3.2 GHz to 5.6 GHz. The measured return-loss results show a slight difference in the frequency range. This difference is primarily due the fabrication process and slight alteration in the dielectric constant of the substrate.

Similarly, the simulation and measured isolation results between the antenna elements are shown in Figure 8B, and isolation between the antenna elements are greater than 16 dB throughout the expected frequency, which demonstrates that all the four antenna elements work independently. A sequence of parameter analyses is presented on the proposed MIMO antenna system to understand the process of the design principle. In Figure 5, the purpose of the rectangular slot at the ground plane is studied with S_{11} and S_{12} measurements by etching with and without the slot in the ground layer. The use of slot C in the ground not only improves the isolation to −22 dB but also enhances the frequency bandwidth. As slot C in the ground plane increases, the frequency bandwidth of the MIMO antenna gradually increases with the frequency range gradually moving back from 4.3 GHz to

3.3 GHz, and similarly the bandwidth also increases from 0.4 GHz to 2.5 GHz, as shown in Figure 9A. Similarly, to understand the effects of using slot E etched in the ground plane, the dimensions of slot E are adjusted from 2 mm to 7 mm while maintaining all other parameter values unchanged. It can be seen in Figure 9B, that the return-loss S_{11} gradually decreases to -10 dB covering the entire frequency range from 3.3 GHz to 5.8 GHz.

Figure 10B shows the working principle of the proposed four-port antenna with the surface current distribution for different frequency bands. This indicates that with the proposed antenna design, the surface current almost does not transfer between the nearby antenna elements at 3.4 GHz, 4.8 GHz, and 5 GHz. This feature assures good isolation between the antenna elements.

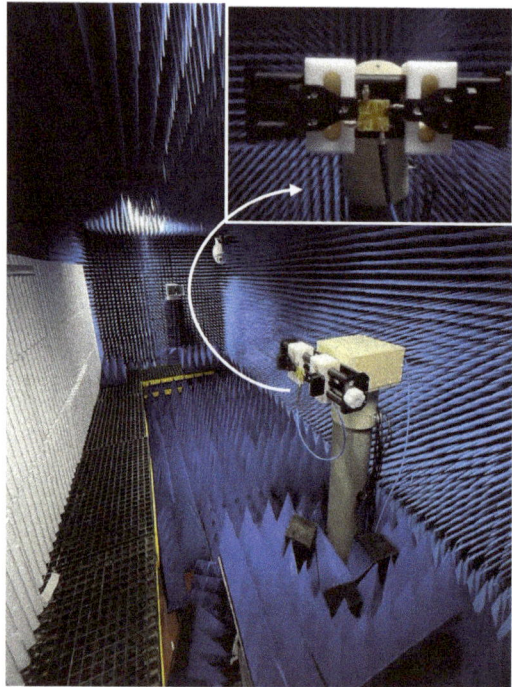

Figure 7. MIMO antenna measurement setup in an anechoic chamber.

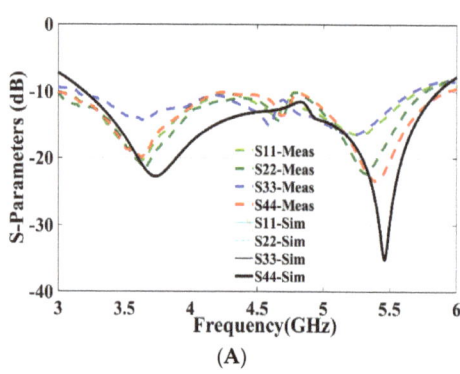
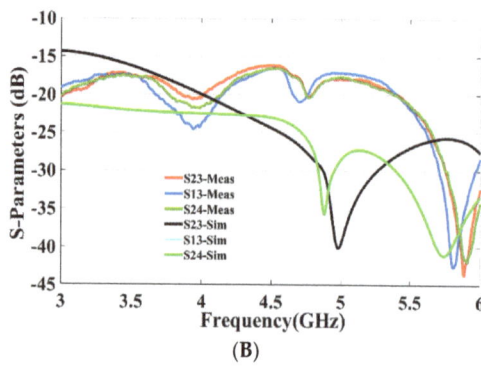

Figure 8. Simulated and measured results of the proposed antenna: (**A**) S_{11}, S_{22}, S_{33}, S_{44} and (**B**) S_{13}, S_{23}, S_{23}, S_{24}.

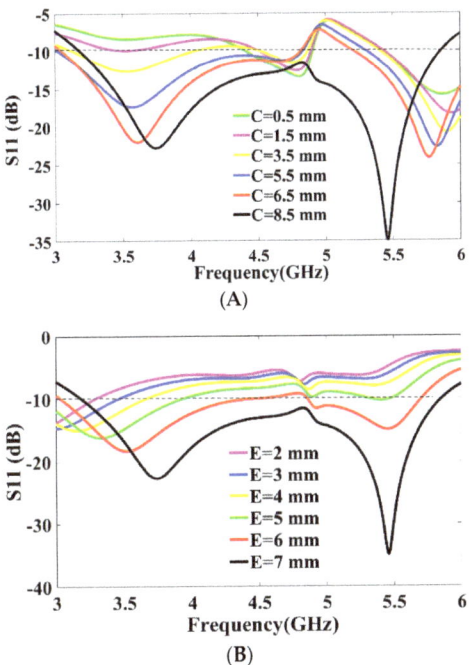

Figure 9. Parametric Analysis: (**A**) S_{11} [dB] vs. Frequency [GHz] for slot C parameter values, (**B**) S_{11} [dB] vs. Frequency [GHz] for slot E parameter values.

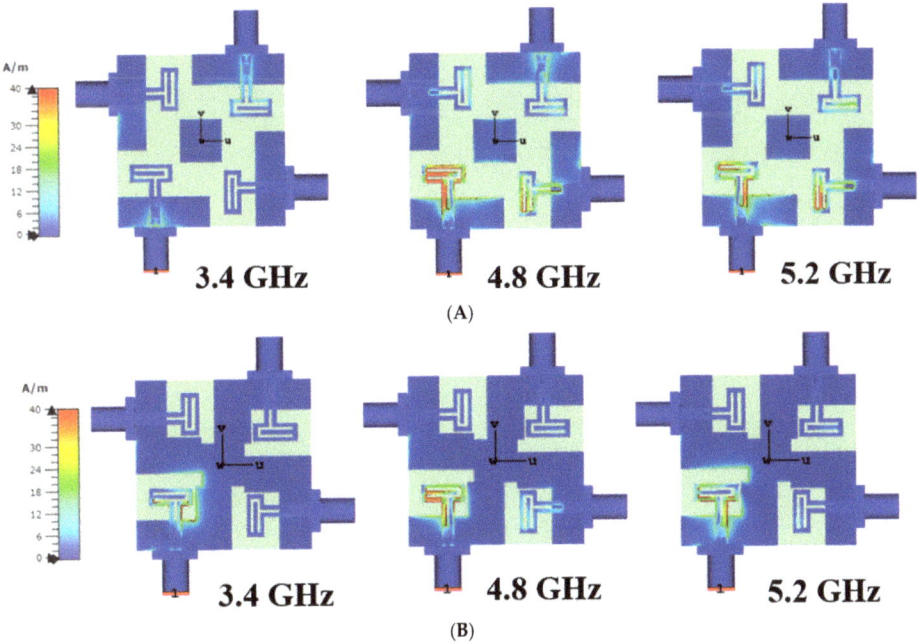

Figure 10. Current distributions in 3.4 GHz, 4.8 GHz and 5.2 GHz. (**A**) Step 2 design antenna; (**B**) final proposed antenna.

Figure 11A–D depicts the simulated and measured 2D YZ-plane and XZ-plane radiation patterns with port 1 excited at 3.4 GHz and 4.8 GHz. The other ports are connected to the 50–ohm match load. It is obvious that the radiation patterns of both the simulated and measured are similar while port 1 is excited and are radiating omnidirectionally. The peak gain achieved by all the four ports lies between 2.4 to 4.9 dBi over the entire operational bandwidth and the average radiation efficiency obtained is 93%. From Figure 11A,B, it can be seen that the maximum measured gain of 2.6 dBi is achieved in the YZ and XZ planes. Similarly, a maximum measured gain of 4 dBi at 4.8 GHz is achieved, as shown in Figure 11C,D.

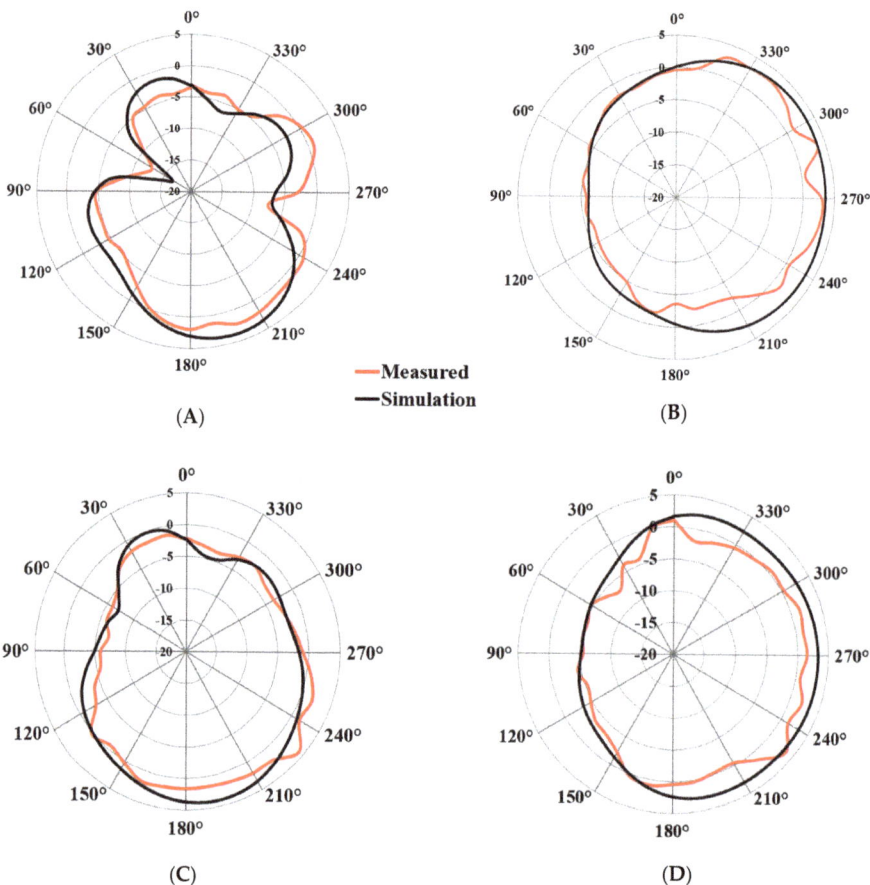

Figure 11. Simulation and measured radiation pattern for port 1: (**A**) 3.4 GHz at YZ plane, (**B**) 3.4 GHz at XZ plane, (**C**) 4.8 GHz at YZ plane, and (**D**) 4.8 GHz at XZ plane.

The ECC and the diversity gain (DG) are important parameters to assess the performance of the MIMO system. The mutual coupling and return loss at the ports can be used to determine ECC, which helps to find the diversity performance of the MIMO antennae [23,24], and is given in Equation (1):

$$|\rho_e(i,j,N)| = \frac{\left|\sum_{n=1}^{N} S_{i,n}^* S_{n,j}\right|}{\sqrt{\left|\Pi_{k(=i,j)}\left[1 - \sum_{n=1}^{N} S_{i,n}^* S_{n,k}\right]\right|}} \qquad (1)$$

The correlation can also be measured from MIMO antenna's far-field radiation patterns [25], as given in Equation (2):

$$ECC = \frac{\left| \int \int_0^4 [E_i(\theta,\phi) * E_j(\theta,\phi)] d\Omega \right|^2}{\int \int_0^4 |E_i(\theta,\phi)|^2 d\Omega \int \int_0^4 |E_j(\theta,\phi)|^2 d\Omega} \quad (2)$$

where i and j are the antenna elements and N is the number of antennae. $E_i(\theta,\phi)$ and $E_j(\theta,\phi)$ are the three-dimensional radiation patterns of ith and jth antenna and Ω is the solid angle. The acceptable and standard value of ECC should be less than 0.5 for portable devices. Similarly, the antenna DG is a well-known performance parameter used to verify the efficacy of the diversity [26]. It can be defined as the ratio of rise in SNR of mixed signals from multiple antennae to the SNR from a single antenna in the system. The DG can be calculated using Equation (3):

$$DG = \sqrt[10]{1 - |ECC|^2} \quad (3)$$

It can be observed that the ECC is less than 0.005 and DG is greater than 9.9 dB in the 3.4 to 6.5 GHz frequency band, as shown in Figure 12. This signifies good diversity performance and shows good performance results in the achieved frequency band. Table 2 provides a comparison between the proposed wideband MIMO antenna and other antenna designs [27–34] found in the literature. This comparison clearly indicates that the proposed antenna design is exceedingly competitive with other designs discussed in the literature in terms of impedance bandwidth, size, and isolation, along with good values of ECC and diversity gain.

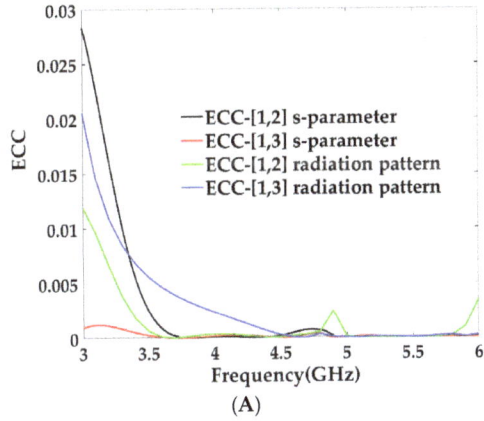
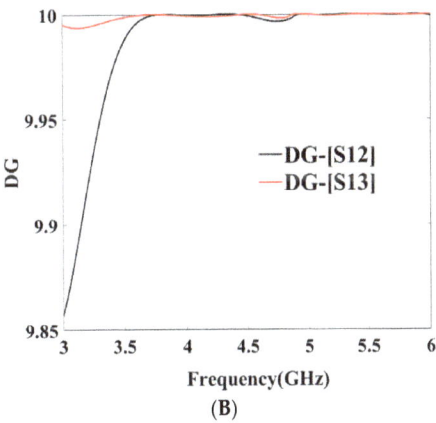

Figure 12. (A) Envelope correlation coefficient [dB] vs. frequency [GHz] (B) diversity gain [dB] vs. frequency [GHz].

Table 2. Comparison between the proposed antenna and other antenna designs in the literature.

Reference/Year	Isolation (dB)	Bandwidth (GHz)	ECC	Isolation/Diversity Technique	Impedance Bandwidth	Total Antenna Size(mm)	Inter-Element Spacing	Common Ground	Number of Ports
[27]/[2020]	19	4.3–6.5	0.004	Orthogonal placement	40%	50 × 50	~0.18λ	No	4
[28]/[2019]	15	2.4, 5.2 and 5.8	0.5	Orthogonal placement	-	52 × 50	~0.10λ	No	4
[29]/[2019]	15	3–10.74	0.1	Parasitic T-shaped strip	112%	81 × 87	~0.77λ	No	4

Table 2. Cont.

Reference/Year	Isolation (dB)	Bandwidth (GHz)	ECC	Isolation/Diversity Technique	Impedance Bandwidth	Total Antenna Size(mm)	Inter-Element Spacing	Common Ground	Number of Ports
[30]/[2019]	17	4.58–6.37	0.05	Parasitic C-shaped	32%	40 × 36	~0.20λ	No	4
[31]/[2019]	15	5.1–5.7	0.05	DGS/decoupling network	11%	50 × 27	~0.16λ	No	4
[32]/[2019]	13	3.3–4.2	0.06	Slots/stubs	24%	42 × 42	~0.15λ	Yes	4
[33]/[2022]	20	3.2–5.7	0.002	EBG	56%	46 × 46	~0.3λ	Yes	4
[34]/[2017]	15	2.3–3.2 and 5.4–5.6	0.05	Polarization diversity/SRR	36%	40 × 40	~0.18λ	Yes	4
This work/[2022]	15	3.2–5.5	0.005	Polarization diversity	54%	40 × 40	~0.17λ	Yes	4

4. Conclusions

A four-port fabricated compact MIMO antenna with interconnected ground plane and simple decoupling structure is proposed and developed covering different sub-6 GHz bands, including 5G and Wi-Fi bands. The measured results show that the impedance bandwidth of 54% (3.2–5.6 GHz) and approximate peak gain of 2.4 to 4.9 dBi over the entire operational bandwidth is achieved. Simple decoupling structure provided good, measured isolation results, better than 16 dB, for the proposed four-port MIMO antenna system, even though the antenna elements are placed close to each other. Furthermore, the measured results and the radiation patterns ensure that the fabricated MIMO antenna system provides a good solution for compact sub-6 GHz MIMO portable devices with diversity performance and for IoT devices.

Author Contributions: Conceptualization, N.S. and H.T.C.; methodology, N.S. and T.K.G.; software, N.S.; validation, H.T.C., B.A.K. and S.K.; formal analysis, H.T.C., N.S. and S.K.; investigation, N.S. and S.K.; resources, S.K. and T.K.G.; data curation, N.S. and H.T.C.; writing—original draft preparation, N.S.; writing—review and editing, H.T.C., B.A.K., S.K. and T.K.G.; visualization, T.K.G.; supervision, H.T.C. and S.K.; project administration, S.K., B.A.K. and T.K.G.; funding acquisition, S.K. and T.K.G. All authors have read and agreed to the published version of the manuscript.

Funding: The authors would like to thank the support of Telekom Malaysia Berhad (TM) under project MMUE/220012 and the Universiti Teknologi Malaysia—grants 03M81 and 4C599.

Institutional Review Board Statement: Not applicable.

Informed Consent Statement: Not applicable.

Data Availability Statement: Not applicable.

Conflicts of Interest: The authors declare no conflict of interest.

References

1. Liaskos, C.; Mamatas, L.; Pourdamghani, A.; Tsioliaridou, A.; Ioannidis, S.; Pitsillides, A.; Akylidiz, I.F. Software-Defined Reconfigurable Intelligent Surfaces: From Theory to End-to-End Implementation. *Proc. IEEE* **2022**, *110*, 1466–1493. [CrossRef]
2. Shafique, K.; Khawaha, B.A.; Sabir, F.; Qazi, S.; Mustaqim, M. Internet of things (IoT) for next-generation smart systems: A review of current challenges, future trends and prospects for emerging 5G-IoT scenarios. *IEEE Access* **2020**, *8*, 23022–23040. [CrossRef]
3. Hua, Q.; Huang, Y.; Alieldin, A.; Song, C.; Jia, T.; Zhu, X. A Dual-Band Dual-Polarized Base Station Antenna Using a Novel Feeding Structure for 5G Communications. *IEEE Access* **2020**, *8*, 63710–63717. [CrossRef]
4. Ban, Y.-L.; Li, C.; Wu, G.; Wong, K.L. 4G/5G multiple antennas for future multi-mode smartphone applications. *IEEE Access* **2016**, *4*, 2981–2988. [CrossRef]
5. Guo, J.; Cui, L.; Li, C.; Sun, B. Side-edge frame printed eight-port dual-band antenna array for 5G smartphone applications. *IEEE Trans. Antennas Propag.* **2018**, *66*, 7412–7417. [CrossRef]
6. Han, C.-Z.; Xiao, L.; Chen, Z.; Yuan, T. Co-Located Self-Neutralized Handset Antenna Pairs with Complementary Radiation Patterns for 5G MIMO Applications. *IEEE Access* **2020**, *8*, 73151–73163. [CrossRef]
7. Hu, W.; Liu, X.; Gao, S.; Wen, L.H.; Qian, L.; Feng, T.; Liu, Y. Dual-band ten-element MIMO array based on dual-mode IFAs for 5G terminal applications. *IEEE Access* **2019**, *7*, 178476–178485. [CrossRef]

8. Parchin, N.O.; Al-Yasir, Y.I.A.; Ali, A.H.; Elfergani, I.; Noras, J.; Rodriguez, J. Eight-element dual-polarized MIMO slot antenna system for 5G smartphone applications. *IEEE Access* **2019**, *7*, 15612–15622. [CrossRef]
9. Ren, Z.; Zhao, A. Dual-band MIMO antenna with compact self-decoupled antenna pairs for 5G mobile applications. *IEEE Access* **2019**, *7*, 82288–82296. [CrossRef]
10. Sun, L.; Feng, H.; Li, Y.; Zhang, Z. Compact 5G MIMO mobile phone antennas with tightly arranged orthogonal-mode pairs. *IEEE Trans. Antennas Propag.* **2018**, *66*, 6364–6369. [CrossRef]
11. Wang, H.; Zhang, R.; Luo, Y.; Yang, G. Compact Eight-Element Antenna Array for Triple-Band MIMO Operation in 5G Mobile Terminals. *IEEE Access* **2020**, *8*, 19433–19449. [CrossRef]
12. Zhao, A.; Ren, Z. Size reduction of self-isolated MIMO antenna system for 5G mobile phone applications. *IEEE Antennas Wirel. Propag. Lett.* **2018**, *18*, 152–156. [CrossRef]
13. Nadeem, I.; Choi, D.-Y. Study on mutual coupling reduction technique for MIMO antennas. *IEEE Access* **2018**, *7*, 563–586. [CrossRef]
14. Ding, K.; Gao, C.; Qu, D.; Yin, Q. Compact broadband MIMO antenna with parasitic strip. *IEEE Antennas Wirel. Propag. Lett.* **2017**, *16*, 2349–2353. [CrossRef]
15. Ghalib, A.; Sharawi, M.S. TCM analysis of defected ground structures for MIMO antenna designs in mobile terminals. *IEEE Access* **2017**, *5*, 19680–19692. [CrossRef]
16. Nandi, S.; Mohan, A. A compact dual-band MIMO slot antenna for WLAN applications. *IEEE Antennas Wirel. Propag. Lett.* **2017**, *16*, 2457–2460. [CrossRef]
17. Su, S.-W.; Lee, C.-T.; Chang, F.-S. Printed MIMO-antenna system using neutralization-line technique for wireless USB-dongle applications. *IEEE Trans. Antennas Propag.* **2011**, *60*, 456–463. [CrossRef]
18. Li, M.-Y.; Xu, Z.Q.; Ban, Y.L.; Sim, C.Y.D.; Yu, Z.F. Eight-port orthogonally dual-polarised MIMO antennas using loop structures for 5G smartphone. *IET Microw. Antennas Propag.* **2017**, *11*, 1810–1816. [CrossRef]
19. Sun, L.; Li, Y.; Zhang, Z.; Feng, Z. Wideband 5G MIMO antenna with integrated orthogonal-mode dual-antenna pairs for metal-rimmed smartphones. *IEEE Trans. Antennas Propag.* **2019**, *68*, 2494–2503. [CrossRef]
20. Xu, H.; Zhou, H.; Gao, S.; Wang, H.; Cheng, Y. Multimode decoupling technique with independent tuning characteristic for mobile terminals. *IEEE Trans. Antennas Propag.* **2017**, *65*, 6739–6751. [CrossRef]
21. Jiang, W.; Cui, Y.; Liu, B.; Hu, W.; Xi, Y. A dual-band MIMO antenna with enhanced isolation for 5G smartphone applications. *IEEE Access* **2019**, *7*, 112554–112563. [CrossRef]
22. Liu, D.Q.; Zhang, M.; Luo, H.J.; Wen, H.L.; Wang, J. Dual-band platform-free PIFA for 5G MIMO application of mobile devices. *IEEE Trans. Antennas Propag.* **2018**, *66*, 6328–6333. [CrossRef]
23. Votis, C.; Tatsis, G.; Kostarakis, P. Envelope correlation parameter measurements in a MIMO antenna array configuration. *Int. J. Commun. Netw. Syst. Sci.* **2010**, *3*, 350. [CrossRef]
24. Abbas, A.; Hussain, N.; Sufian, M.A.; Jung, J.; Park, S.M.; Kim, N. Isolation and Gain Improvement of a Rectangular Notch UWB-MIMO Antenna. *Sensors* **2022**, *22*, 1460. [CrossRef] [PubMed]
25. Sufian, M.A.; Hussain, N.; Askari, H.; Park, S.G.; Shin, K.S.; Kim, N. Isolation enhancement of a metasurface-based MIMO antenna using slots and shorting pins. *IEEE Access* **2021**, *9*, 73533–73543. [CrossRef]
26. Liu, X.; Amin, M.; Liang, J. Wideband MIMO antenna with enhanced isolation for wireless communication application. *IEICE Electron. Express* **2018**, *15*, 20180948. [CrossRef]
27. Mohammad Saadh, A.; Khangarot, S.; Sravan, B.V.; Aluru, N.; Ramaswamy, P.; Ali, T.; Pai, M.M. A compact four-element MIMO antenna for WLAN/WiMAX/satellite applications. *Int. J. Commun. Syst.* **2020**, *33*, e4506. [CrossRef]
28. Roy, S.; Ghosh, S.; Chakarborty, U. Compact dual wide-band four/eight elements MIMO antenna for WLAN applications. *Int. J. RF Microw. Comput. Aided Eng.* **2019**, *29*, e21749. [CrossRef]
29. Srivastava, K.; Kumar, A.; Kanaujia, B.K.; Dwari, S.; Kumar, S. A CPW-fed UWB MIMO antenna with integrated GSM band and dual band notches. *Int. J. RF Microw. Comput. Aided Eng.* **2019**, *29*, e21433. [CrossRef]
30. Nie, N.S.; Yang, X.S.; Wang, B.Z. A compact four-element multiple-input-multiple-output antenna with enhanced gain and bandwidth. *Microw. Opt. Technol. Lett.* **2019**, *61*, 1828–1834. [CrossRef]
31. Nandi, S.; Mohan, A. A self-diplexing MIMO antenna for WLAN applications. *Microw. Opt. Technol. Lett.* **2019**, *61*, 239–244. [CrossRef]
32. Barani, I.R.R.; Wong, K.L.; Zhang, Y.X.; Li, W.Y. Low-Profile Wideband Conjoined Open-Slot Antennas Fed by Grounded Coplanar Waveguides for 4 × 4 5 G MIMO Operation. *IEEE Trans. Antennas Propag.* **2019**, *68*, 2646–2657. [CrossRef]
33. Megahed, A.A.; Abdelazim, M.; Abdelhay, E.H.; Soliman, H.Y. Sub-6 GHz highly isolated wideband MIMO antenna arrays. *IEEE Access* **2022**, *10*, 19875–19889. [CrossRef]
34. Sarkar, D.; Srivastava, K. Compact four-element SRR-loaded dual-band MIMO antenna for WLAN/WiMAX/WiFi/4G-LTE and 5G applications. *Electron. Lett.* **2017**, *53*, 1623–1624. [CrossRef]

Article

A Dual-Band Eight-Element MIMO Antenna Array for Future Ultrathin Mobile Terminals

Chuanba Zhang, Zhuoni Chen, Xiaojing Shi, Qichao Yang, Guiting Dong, Xuanhe Wei and Gui Liu *

College of Electrical and Electronic Engineering, Wenzhou University, Wenzhou 325035, China
* Correspondence: gliu@wzu.edu.cn

Abstract: An ultrathin dual-band eight-element multiple input–multiple output (MIMO) antenna operating in fifth-generation (5G) 3.4–3.6 GHz and 4.8–5 GHz frequency bands for future ultrathin smartphones is proposed in this paper. The size of a single antenna unit is 9×4.2 mm^2 ($0.105 \lambda \times 0.05 \lambda$, λ equals the free-space wavelength of 3.5 GHz). Eight antenna units are structured symmetrically along with two sideboards. Two decoupling branches (DB1 and DB2) are employed to weaken the mutual coupling between Ant. 1 and Ant. 2 and between Ant. 2 and Ant. 3, respectively. The measured −10 dB impedance bands are 3.38–3.82 GHz and 4.75–5.13 GHz, which can entirely contain the desired bands. Measured isolation larger than 14.5 dB and 15 dB is obtained in the first and second resonant modes, respectively. Remarkable consistency between the simulated and measured results can be achieved. Several indicators, such as the envelope correlation coefficient (ECC), diversity gain (DG), total active reflection coefficient (TARC), and multiplexing efficiency (ME), have been presented to assess the MIMO performance of the designed antenna.

Keywords: 5G; multiple input–multiple output (MIMO); ultrathin; smartphone; dual-band

Citation: Zhang, C.; Chen, Z.; Shi, X.; Yang, Q.; Dong, G.; Wei, X.; Liu, G. A Dual-Band Eight-Element MIMO Antenna Array for Future Ultrathin Mobile Terminals. *Micromachines* **2022**, *13*, 1267. https://doi.org/10.3390/mi13081267

Academic Editors: Lu Zhang, Xiaodan Pang and Prakash Pitchappa

Received: 10 July 2022
Accepted: 1 August 2022
Published: 6 August 2022

Publisher's Note: MDPI stays neutral with regard to jurisdictional claims in published maps and institutional affiliations.

Copyright: © 2022 by the authors. Licensee MDPI, Basel, Switzerland. This article is an open access article distributed under the terms and conditions of the Creative Commons Attribution (CC BY) license (https://creativecommons.org/licenses/by/4.0/).

1. Introduction

Fifth-generation (5G) communication is burgeoning, demanding wireless devices with a transmission data rate as high as possible. Multiple input–multiple output (MIMO) technology possesses promising application prospects in improving the data rate. Recently, many sub-6 GHz 5G smartphone MIMO antennas have been developed [1–16], such as four-element MIMO antennas [2–5], eight-port smartphone antennas [6–10], and even twelve-element MIMO antennas [11,12]. One nonnegligible challenge encountered during the design process is the method to effectively weaken the mutual electromagnetic coupling between antenna elements in a MIMO antenna array. However, numerous decoupling mechanisms have been put forward, such as polarization diversity [9], defected ground structure (DGS) [12,13], decoupling branches [14], neutralization lines [15], and orthogonal mode [16]. More attention still needs to be focused on the decoupling design in the MIMO antenna array.

The usual height of the lateral side frame of a conventional smartphone's antenna [2,3,6,7] is 7 mm, which is not conducive to implementing future ultrathin smartphones. Some low-profile MIMO antennas for the 5G handsets have been proposed recently. In [17], a compact four-element MIMO antenna pair for 5G mobile was presented, integrating two antenna elements at a close distance of 1.2 mm. The designed antenna pair resonated precisely at 3.5 GHz. One more worthy mention is that the overall volume of the MIMO system was $150 \times 73 \times 6$ mm^3, which realized a 1 mm reduction in the height of the lateral side frame. Another self-decoupled four-element antenna pair [18] functioning in the 3.5 GHz band (3.4–3.6 GHz) with the same height of 6 mm has been presented, and the mutual coupling of the antenna pair was decreased to 16.5 dB. In [19], a low-profile, high-isolation eight-port MIMO antenna for the 5G handset was presented, and the height of the lateral sideboard was 5.3 mm.

This paper presents an ultrathin eight-port MIMO antenna working at 5G 3.4–3.6 GHz and 4.8–5 GHz frequency bands. The integral volume of the proposed antenna is only $145 \times 70 \times 5$ mm^3, which is thinner than other published 5G smartphone antennas. Eight antenna elements are manufactured along the inner face of two sideboards. Two decoupling branches (DB1 and DB2) are employed to attenuate the mutual coupling. The proposed antenna is fabricated and measured. The measured −10 dB impedance bands are 3.38–3.82 GHz and 4.75–5.13 GHz, which can fully contain two target bands. A measured lowest isolation (14.5 dB) emerged in S_{23} around 3.5 GHz. DHM mode is provided to assess practical application ability. ECC, DG, TARC, and ME are calculated to evaluate diversity performance.

2. Antenna Structure

The overall view and lateral perspective of the proposed antenna array are shown in Figure 1. Eight antenna elements are printed along the inner side of two sideboards with a size of $145 \times 4.2 \times 0.8$ mm^3, which are constructed perpendicularly to the system board. The size of the system board is $145 \times 70 \times 0.8$ mm^3. The sideboards and system board substrate are an FR4 substrate with loss tangent = 0.02 and relative permittivity = 4.4. The height of the whole smartphone is only 5 mm, since the sideboards are placed on the system board. A 2 mm-wide microstrip line feeds each element through an SMA connector via the hole from the bottom of the system board. The designed DBs are separately printed on the inner face and upper side of the sideboard and system board, which are welded together. A ground plane (145×70 mm^2) with two rectangular ground clearances (145×3.5 mm^2) is fabricated on the bottom of the system board. The dimensions of DB1(2) and the detailed construction of a fundamental antenna element are illustrated in Figure 1c,d. Parameters that affect antenna performance are described as variables rather than a fixed value. Notably, the values of $S1$ of DB1 and DB2 are 11 mm and 9 mm, respectively.

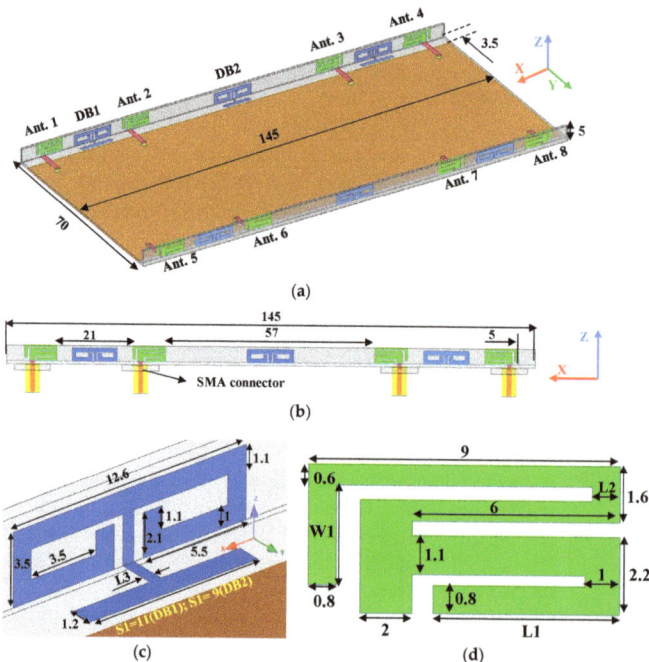

Figure 1. Perspectives of the proposed antenna. (**a**) Overall view, (**b**) side view, (**c**) dimensions of DB1 and DB2, and (**d**) detailed structure of an antenna element. (All values are in millimeters).

3. Working Mechanism and Application Scenario

In this section, the design evolution steps are first presented to understand the operating mechanism better. Consequently, the role of the DBs is analyzed. The third part shows the current vector distribution of the proposed antenna at 3.5 GHz and 4.9 GHz when DB1 is utilized or not, and some variables are selected to be analyzed. The last portion of this section provides a scenario where this device is held in dual-hand mode (DHM).

3.1. Design Procedure

This section presents a precise design evaluation of the proposed antenna. Figure 2a gives the four structures during the design process. The first structure is a single rectangular plane with a small open-ended L-shaped slot. It can be seen from Figure 2c that an obvious resonant mode around 4.3 GHz of Ant. 1 and Ant. 2 is obtained. However, the port impedance matching needs to be optimized. As shown in Figure 2b, the simulated normalized port impedance curve is far away from the center point of the Smith chart. Another small rectangular slot and ground clearance (145 × 3.5 mm^2) are cut from the antenna element and grounding plane in the second structure. A T-shaped strip is introduced to diminish the mutual coupling between Ant. 1 and Ant. 2. It can be distinctly observed from Figure 2c that two resonant modes (around 3.6 GHz and 5.5 GHz) are excited. Little frequency offset between S_{11} and S_{22} occurs because of the two elements' different locations. The mutual coupling S_{12} of the second structure is 10 dB and 13 dB in the lower and higher bands, respectively.

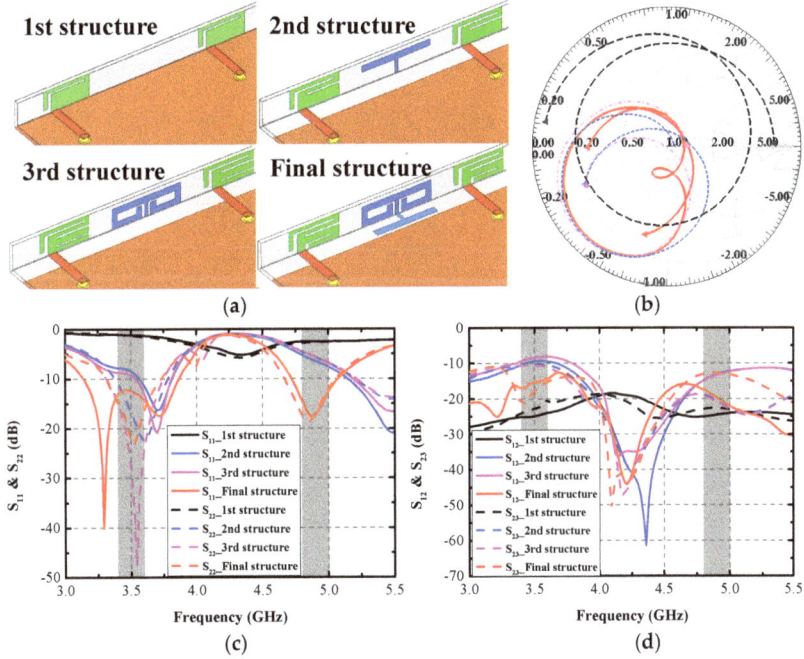

Figure 2. (a) Design evolution, (b) simulated Smith chart of S_{11}, (c) simulated S_{11} and S_{22}, (d) simulated S_{12} and S_{23}.

Furthermore, a significant enhancement in the normalized impedance matching condition is acquired, which can be observed in Figure 2b. In the third structure, a small rectangular slot is cut from the antenna element of the second structure. An extra two C-shaped strips are added to the upper terminals of the aforementioned T-shaped decoupling branch. The simulated S_{22} obtains a good matching condition at 3.53 GHz. The lowest isolation of 8 dB occurs at 3.53 GHz, and little isolation promotion at 5.5 GHz is realized. The final antenna element structure is produced by cutting a small rectangular slot in the feedline side of the lowest rectangular strip and reconnecting it on the other side. At the same time, another newly introduced horizontal T-shaped branch is connected to the existing decoupling branch as the third structure. The final design of an antenna element and DB1 are generated in Figure 2a. The simulated S_{11} and S_{22} can entirely contain the target bands, and the isolation S_{21} is also lifted to 15 dB and 18 dB across 3.4–3.6 GHz and 4.8–5 GHz, respectively. Table 1 lists the simulated normalized impedance at each design stage at 3.5 GHz and 4.9 GHz. The final structure obtains better impedance matching performance than the former three design stages.

Table 1. The simulated normalized impedance at two resonant frequencies of each design stage.

Design Evolution	3.5 GHz	4.9 GHz
1st structure	0.094 + 0.518i	0.2946 + 0.9528i
2nd structure	0.5085 + 0.0955i	0.3626 + 0.2061i
3rd structure	0.6245 + 0.0824i	0.3102 + 0.2382i
Final structure	1.08–0.3846i (3.35 GHz) 0.7507–0.3014i (3.5 GHz)	1.3667 + 0.0222i

3.2. Study of the Role of the DBs

This section presents the simulated results with/without DBs. As shown in Figure 3a, when there is no DB1, the simulated S_{11} and S_{22} can still cover the desired bands. However, the first resonant frequency of Ant. 1 moves to 3.6 GHz, while the other operation band causes little influence. Figure 3b illustrates the simulated isolation curves S_{12} and S_{23} with/without DB1 and DB2. The utilization of the DBs can effectively attenuate the mutual coupling at 3.5 GHz, while there is little impact on the mutual coupling at 4.9 GHz, as shown in Figure 4. The worst simulated isolation (14 dB) appeared at S_{12}. Figure 3c portrays the simulated S_{23} with various values of *S1*. Relatively low isolation (8 dB) at 3.5 GHz was obtained when no DB was used. After the DB1 is constructed between Ant. 2 and Ant. 3, a distinctly improving trend occurred to S_{23}, as depicted in Figure 3c, but it was still insufficient. By adjusting the length of *S1*, isolation performance can be improved. When the value of *S1* is 9 mm, the simulated S_{23} satisfies the requirement of 15 dB within the desired bands at 3.5 GHz. When the value of *S1* decreases to 7 mm, some deterioration happens to S_{23}, as shown in Figure 3c. The final optimized value of *S1* is 9 mm.

3.3. Current Distribution and Parametric Analysis

The simulated current distribution of Ant. 1 and Ant. 2 at two operating frequencies when the DB1 is adopted or not are portrayed in Figure 4. When Ant. 1 is excited at 3.5 GHz, the strongest current density is allocated over the upper L-shaped slot of Ant. 1 and the inner edges of slots of the lateral section of DB1. The introduction of DB1 significantly decreases the current density coupled in Ant. 2 when Ant. 1 is excited at 3.5 GHz. When Ant. 2 is excited at 4.9 GHz, the maximum current spread around the middle rectangular slot of Ant. 2. There is no significant difference in the coupling current of distribution of Ant. 1 when DB1 is applied or not. The utilization of DB1 powerfully absorbs the mutual magnetic coupling existing between Ant. 1 and Ant. 2 at 3.5 GHz, hence enhancing the isolation. The slight improvement resulting from the DB1 is realized upon S_{12} at 4.9 GHz, which is also consistent with the simulated curves in Figure 3b.

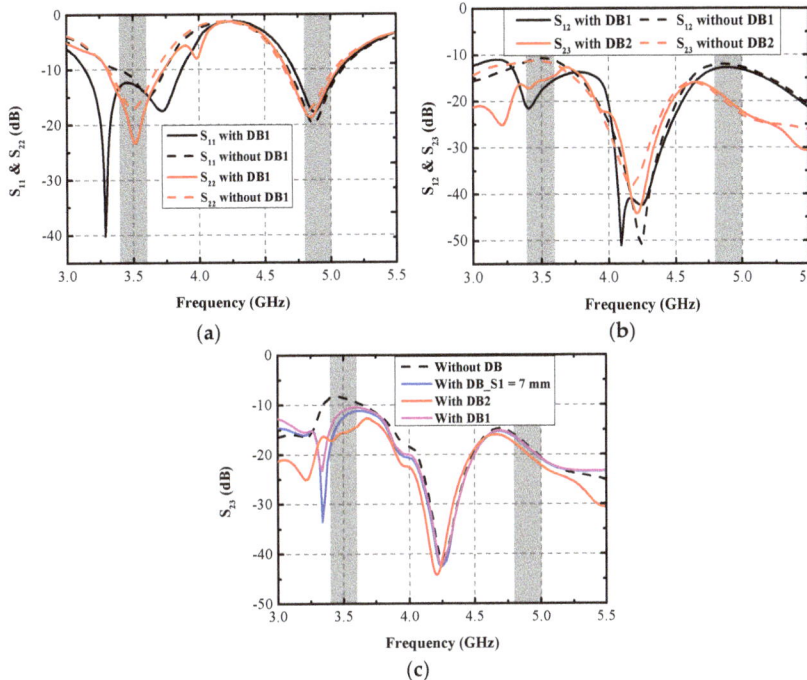

Figure 3. Simulated (**a**) S_{11} and S_{22} and (**b**) $S_{12/23}$ with/without DBs, (**c**) S_{23} with various values of $S1$.

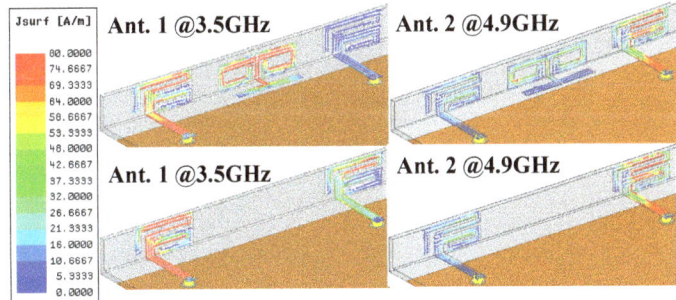

Figure 4. Current distribution at two resonant modes when DB1 is utilized or not.

According to the current distribution, numerous parameters are selected to make a parametric analysis, as shown in Figure 5. Little resonant frequency offset of the latter band arises with the increase in $L1$, and almost no impact is caused on the first operating band. The final value of $L1$ is 5.4 mm. As illustrated in Figure 5b, the variation in $L2$ affects all three resonant points. With the increase in $L2$, the impedance matching condition at 3.5 GHz deteriorates, the middle resonant mode around 3.7 GHz shifts to the higher frequency, and the ultimately optimized length of $L2$ is 0.8 mm. Without influencing the first resonant mode of Ant. 1, the addition of the value of $W1$ contributes a lot to the movement of the other two resonant modes. As illustrated in Figure 5c, the second resonant mode moves to 4 GHz when the value of $W1$ is 2.2 mm, and the final impedance band is not satisfied. The ultimately modified value of $W1$ is 2.8 mm. Parameter $L3$ mainly affects the decoupling performance. A significant difference in the isolation at 3.5 GHz occurs with the varying of L3. When the value of L3 equals 0.6 mm, the simulated S_{21} is separately larger than 15 dB and 14 dB at 3.5 GHz and 4.9 GHz, respectively.

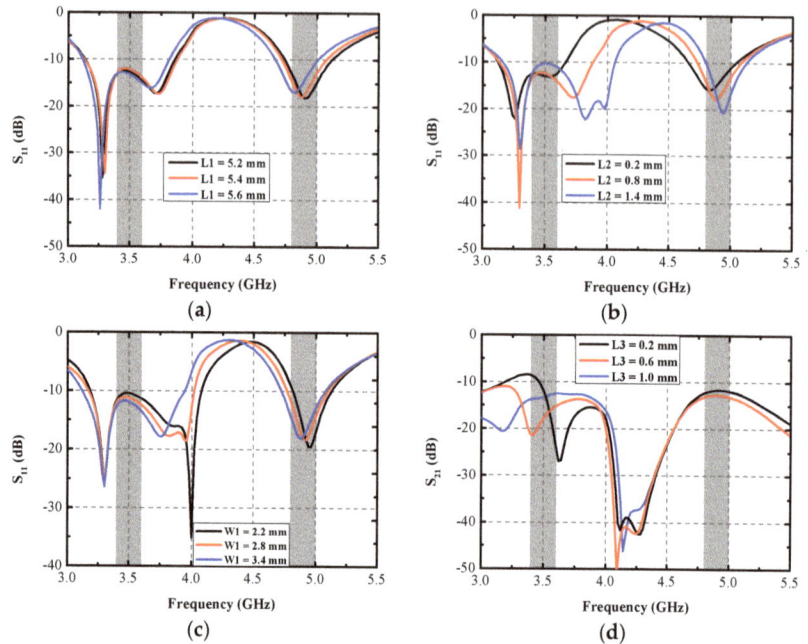

Figure 5. Simulated S_{11} with various (**a**) L1, (**b**) L2, (**c**) W1, and (**d**) S_{21} with different L3.

3.4. Application Scenario

An application scenario of the presented antenna held in dual-hand mode (DHM) is studied to identify the robustness and practicability of the proposed MIMO antenna array. Figure 6 plots the simulated S-parameters in DHM and the −10 dB bandwidth of Ant. 5, with Ant. 8 not being able cover two target bands. The −10 dB impedance matched bandwidth of Ant. 2, Ant. 3, Ant. 6, and Ant. 7 can contain 5G 3.6–3.8 GHz and 4.8–5 GHz frequency bands. The simulated S_{11} and S_{44} can still wholly cover the two desired bands. Figure 6b provides the total radiated power (TRP) of the proposed antenna when Ant. 1, Ant. 2, Ant. 5, and Ant. 6 are separately excited with 1 W input power. The radiating ability of four inner elements (Ant. 2, Ant. 3, Ant. 6, and Ant. 7) are generally better than the other four elements constructed in the corners of the system substrate, which have the closest distance to the hand tissue compared with the inner four elements. Figure 7 presents the proposed antenna's simulated three-dimension (3D) and two-dimension (2D) radiation patterns when Ant. 8 and Ant. 7 are independently excited at 3.5 GHz and 4.9 GHz, respectively. The simulated specific absorption rate (SAR) distribution when Ant. 8 and Ant. 7 are separately excited with 100 mW input power at two resonant modes, as shown in Figure 8. A maximum SAR value of 1.45 W/kg and 1.22 W/kg is acquired at 3.5 GHz and 4.9 GHz, respectively. Both SAR values are lower than the European and American requirements of 2.0 W/kg and 1.6 W/kg.

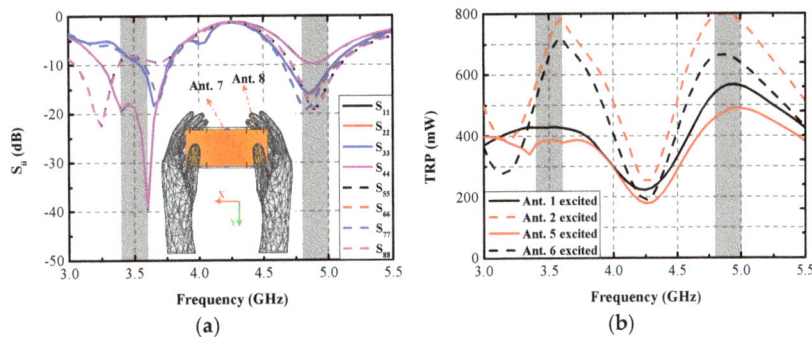

Figure 6. Simulated (**a**) S-parameters and (**b**) TRPs for the presented antenna held in DHM.

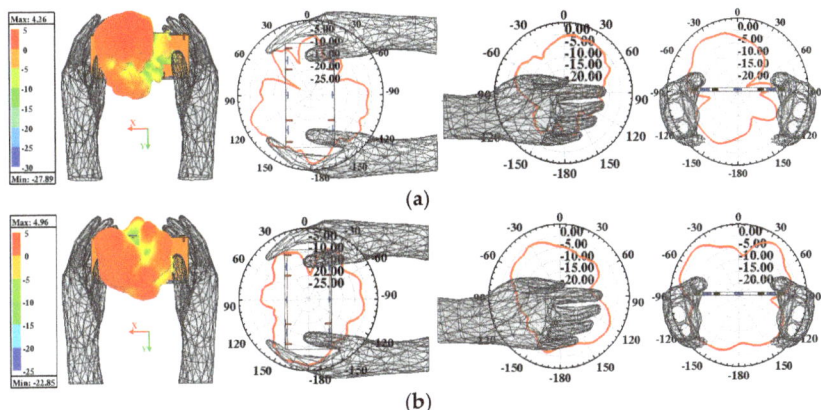

Figure 7. Simulated 3D and 2D patterns when (**a**) Ant. 8 excited at 3.5 GHz and (**b**) Ant. 7 excited at 4.9 GHz.

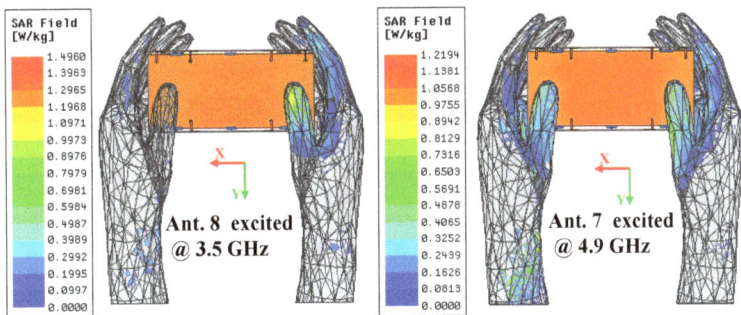

Figure 8. SAR field distribution.

4. Experimental Results

A prototype of the explored antenna was printed and measured to validate the simulated results. Figure 9 presents the photograph of the prototype and test scenarios using a vector network analyzer (VNA: N5224A) and anechoic chamber. In Figure 9a, when Ant. 2 and Ant. 3 are excited, two distinctly resonant modes around 3.5 GHz and 4.9 GHz can be obtained, and excellent uniformity between S_{22} and S_{33} can be observed. Little frequency offset occurs because of the soldering process of the SMA connectors. Figure 9b illustrates the measuring environment of the 2D radiating patterns. Figure 10a,b compare the simulated and measured S-parameters (Sii and Sij, respectively). A slight frequency shift exists between the simulated and measured results, but the measurement can still completely cover the target bands. Measured worst isolation (14.5 dB) of S_{23} appears at around 3.5 GHz. Figure 10c,d present the measured S-parameters of the proposed antenna. All the tested input return loss curves of eight ports can contain the desired bands, and the measured mutual coupling is separately larger than 14.5 dB and 15 dB at 3.5 GHz and 4.9 GHz. Figure 11 provides the simulated and measured gain and radiating efficiency of Ant. 1 and Ant. 2. As shown in Figure 11a, maximum gains of 5 dBi and 4.8 dBi are achieved during the former and latter operating bands, respectively. Radiating efficiency of approximately 60% and 70% is obtained separately at 3.5 GHz and 4.9 GHz, as shown in Figure 11b. The measured and simulated 2D radiating patterns of the proposed antenna are illustrated in Figure 12. The discrepancies between the simulated and measured curves are caused by the soldering process and the installation angle of the antenna when it is tested in the anechoic chamber.

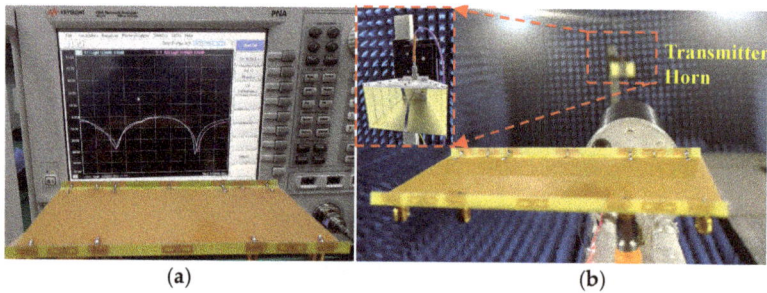

Figure 9. Photograph of the manufactured model of the proposed antenna and the experimental environment (**a**) measured by the VNA and (**b**) measured in the anechoic chamber.

Numerous indicators, including ECC, DG, TARC, and ME, were computed to assess the MIMO performance of the designed MIMO antenna. Figure 13 . The largest measured ECCs of 0.004 and 0.008 are realized across the former and the latter operating modes, respectively. The ECCs are computed from the radiating results based on Formula (1) [18]. The computed DGs, calculated from Formula (2) [19], are better than 9.99 dB and 9.978 dB within the two target bands, respectively. TARC is the definition of the square root of the ratio of total reflected radio-frequency (RF) power to the total incident power. As shown in Figure 14, the TARC curves are calculated by Equation (3) [20], which are well below the −10 dB level within the two desired bands. ME is defined as the power loss of a realistic antenna in achieving a given power capacity compared with an ideal antenna with total percentage radiation efficiency. ME can be expressed by Equation (4) [21]. Figure 15 compares the simulated and measured ME results between Ant. 1 and Ant. 2, and between Ant. 2 and Ant. 3, respectively. Measured ME values of approximately 70% and 75% are obtained at 3.5 GHz and 4.9 GHz, respectively. Remarkable consistency between the simulated and measured ME curves was observed.

$$\text{ECC} = \frac{\left|S_{ii}^*S_{ij} + S_{ji}^*S_{jj}\right|^2}{(1 - |S_{ii}|^2 - |S_{ji}|^2)(1 - |S_{jj}|^2 - |S_{ij}|^2)} \tag{1}$$

$$\text{DG} = 10 \times \sqrt{1 - \text{ECC}^2} \tag{2}$$

$$\text{TARC} = \sqrt{\frac{(S_{11} + S_{12})^2 + (S_{22} + S_{21})^2}{2}} \tag{3}$$

$$\mathrm{ME} = \sqrt{\eta_1 \eta_2 \left(1 - \mathrm{ECC}_{12}^2\right)} \qquad (4)$$

where η_1 and η_2 represent the total efficiency of Ant. 1 and Ant. 2, respectively.

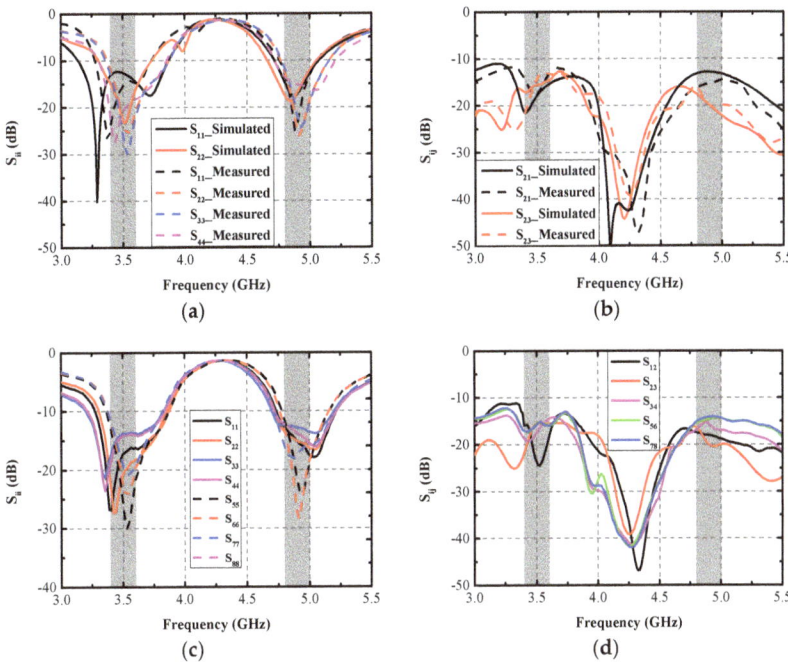

Figure 10. Simulated and measured S-parameters (**a**) S_{ii}, (**b**) S_{ij}, (**c**) measured S_{ii}, and (**d**) measured S_{ij}.

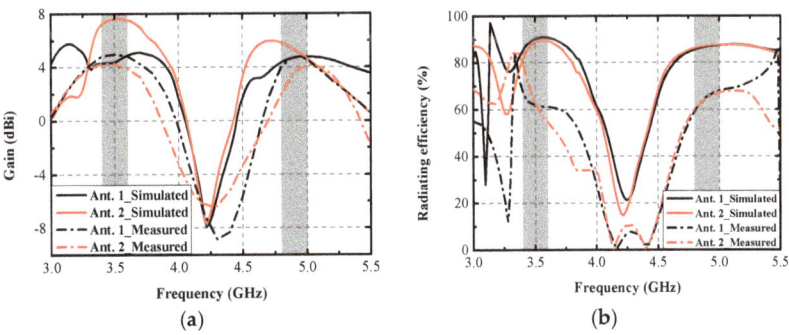

Figure 11. Simulated and measured (**a**) gain and (**b**) radiating efficiency of Ant. 1 and Ant. 2.

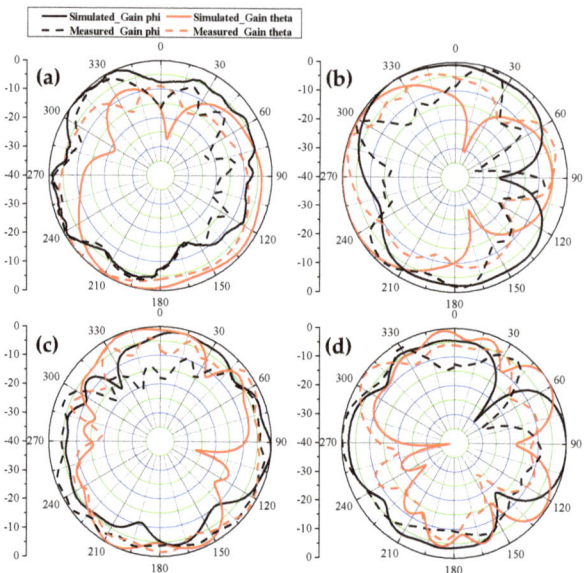

Figure 12. Measured and simulated 2D radiating patterns. (**a**) Ant. 1 is excited at 3.5 GHz, XOY plane, (**b**) Ant. 1 is excited at 3.5 GHz, XOZ plane, (**c**) Ant. 2 is excited at 4.9 GHz, XOY plane, (**d**) Ant. 2 is excited at 4.9 GHz, XOZ plane.

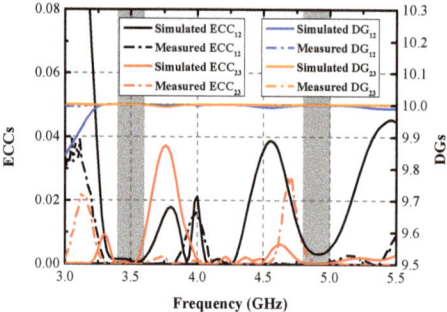

Figure 13. Simulated and measured ECCs and DGs.

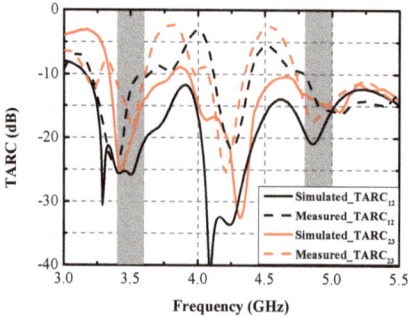

Figure 14. Simulated and measured TARCs.

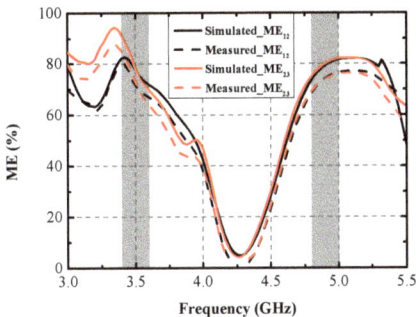

Figure 15. Simulated and measured MEs.

Table 2 presents a performance contrast between the presented antenna and other 5G smartphone antennas exploited in recent years. The primary highlights of the proposed antenna are the lowest lateral sideboard, superior isolation performance, and lower ECCs.

Table 2. Performance contrast between this work and other reported 5G smartphone antennas.

Design	Working Band (GHz)	Total Size (mm³)	Dimension of A Single Element (mm³)	Decoupling Method	Isolation (dB)	ECC
[14]	3.4–3.6 4.8–5 (−6 dB)	150 × 75 × 7	14.8 × 7 × 0.8	Decoupling structure	15.5 19	0.07 0.06
[15]	3.4–3.6 4.8–5 (−6 dB)	150 × 75 × 7	15 × 7 × 0.8	Neutralization line	11.5	0.08
[16]	3.4–3.6 (−10 dB)	150 × 73 × 6	12 × 4.2 × 0.8	Orthogonal Mode	17	0.06
[22]	3.4–3.6 4.8–5 (−10 dB)	150 × 75 × 6	17.4 × 6 × 0.8	Self-isolated	19.1	0.0125
[23]	3.3–3.6 4.8–5 (−10 dB)	150 × 73 × 7	15.5 × 7 × 0.8	Self-isolated	11	0.15
[24]	4.4–5 (−6 dB)	150 × 80 × 0.787	50 × 30 × 0.787	Shorting pins	18	0.24
[25]	3.4–3.6 (−10 dB)	150 × 75 × 5.3	16.1 × 4.5	Self-isolated	20	0.4
This work	3.4–3.6 4.8–5 (−10 dB)	145 × 70 × 5	9 × 4.2 × 0.8	DBs	14.5 15	0.004 0.008

5. Conclusions

An ultrathin eight-port MIMO antenna functioning in 5G 3.4–3.6 GHz and 4.8–5 GHz is presented in this paper. The explored antenna element obtained a minimized dimension of 9×4.2 mm^2, and the overall volume of the MIMO system was only $145 \times 70 \times 5$ mm^3. Two kinds of DBs were employed to attenuate the mutual coupling at 3.5 GHz. Besides the design stages of the proposed antenna, the role performed by the DBs were also studied to gain profound understanding of the decoupling mechanism. An application scenario of DHM was given to evaluate the robustness and practicability of the presented antenna. The measured −10 dB impedance band is able to contain the target bands entirely. The measured worst mutual coupling (14.5 dB) appeared in S_{23} around 3.5 GHz. Maximum radiating efficiency of 60% and 75% were obtained within the first and second bands, respectively. The computed results of indicators, such as the ECC (0.008), DG (9.978), TARC (10 dB), and ME (70%), have proved the excellent MIMO performance of the proposed antenna.

Author Contributions: Conceptualization, C.Z.; methodology, Z.C. and G.D.; investigation, Q.Y. and X.W.; writing—original draft preparation, C.Z.; writing—review and editing, X.S. and G.L.; supervision and funding acquisition, G.L. All authors have read and agreed to the published version of the manuscript.

Funding: This work was partly supported by the National Natural Science Foundation of China under grant 61671330, the Science and Technology Department of Zhejiang Province under grant LGG19F010009, and Wenzhou Municipal Science and Technology Program under grants C20170005 and 2018ZG019.

Institutional Review Board Statement: Not applicable.

Informed Consent Statement: Not applicable.

Data Availability Statement: Not applicable.

Conflicts of Interest: The authors declare no conflict of interest.

References

1. Liu, D.; Luo, H.; Zhang, M.; Wen, H.; Wang, B.; Wang, J. An Extremely Low-Profile Wideband MIMO Antenna for 5G Smartphones. *IEEE Trans. Antennas Propag.* **2019**, *67*, 5772–5780. [CrossRef]
2. Huang, J.; Dong, G.; Cai, Q.; Chen, Z.; Li, L.; Liu, G. Dual-Band MIMO Antenna for 5G/WLAN Mobile Terminals. *Micromachines* **2021**, *12*, 489. [CrossRef] [PubMed]
3. Huang, J.; Dong, G.; Cai, J.; Li, H.; Liu, G. A Quad-Port Dual-Band MIMO Antenna Array for 5G Smartphone Applications. *Electronics* **2021**, *10*, 542. [CrossRef]
4. Ren, Z.; Zhao, A. Dual-Band MIMO Antenna with Compact Self-Decoupled Antenna Pairs for 5G Mobile Applications. *IEEE Access* **2019**, *7*, 82288–82296. [CrossRef]
5. Moses, A.; Moses, N. Compact Self Decoupled MIMO Antenna Pairs Covering 3.4–3.6 GHz Band for 5G Handheld Device Applications. *AEU Int. J. Electron. Commun.* **2021**, *141*, 153971. [CrossRef]
6. Sun, L.; Li, Y.; Zhang, Z. Wideband Decoupling of Integrated Slot Antenna Pairs for 5G Smartphones. *IEEE Trans. Antennas Propag.* **2021**, *69*, 2386–2391. [CrossRef]
7. Wong, K.; Tsai, C.; Lu, J. Two Asymmetrically Mirrored Gap-Coupled Loop Antennas as a Compact Building Block for Eight-Antenna MIMO Array in the Future Smartphone. *IEEE Trans. Antennas Propag.* **2017**, *65*, 1765–1778. [CrossRef]
8. Huang, J.; He, T.; Xi, S.; Yang, Q.; Shi, X.; Liu, G. Eight-port high-isolation antenna array for 3.3–6 GHz handset applications. *AEU Int. J. Electron. Commun.* **2022**, *154*, 154333. [CrossRef]
9. Li, Y.; Sim, C.; Luo, Y.; Yang, G. High-Isolation 3.5 GHz Eight-Antenna MIMO Array Using Balanced Open-Slot Antenna Element for 5G Smartphones. *IEEE Trans. Antennas Propag.* **2019**, *67*, 3820–3830. [CrossRef]
10. Jiang, W.; Liu, B.; Cui, Y.; Hu, W. High-Isolation Eight-Element MIMO Array for 5G Smartphone Applications. *IEEE Access* **2019**, *7*, 34104–34112. [CrossRef]
11. Dong, J.; Wang, S.; Mo, J. Design of a Twelve-Port MIMO Antenna System for Multi-Mode 4G/5G Smartphone Applications Based on Characteristic Mode Analysis. *IEEE Access* **2020**, *8*, 90751–90759. [CrossRef]
12. Yuan, X.; He, W.; Hong, K.; Han, C.; Chen, Z.; Yuan, T. Ultra-Wideband MIMO Antenna System with High Element-Isolation for 5G Smartphone Application. *IEEE Access* **2020**, *8*, 56281–56289. [CrossRef]
13. Dong, G.; Huang, J.; Chen, Z.; Liu, G. A Compact Planar Dual Band Two-Port MIMO Antenna with High Isolation and Efficiency. *Int. J. RF Microw. Comput. Aided Eng.* **2022**, *32*, e23245. [CrossRef]
14. Hu, W. Dual-Band Eight-Element MIMO Array Using Multi-Slot Decoupling Technique for 5G Terminals. *IEEE Access* **2019**, *7*, 153910–153920. [CrossRef]
15. Guo, J.; Cui, L.; Li, C.; Sun, B. Side-Edge Frame Printed Eight-Port Dual-Band Antenna Array for 5G Smartphone Applications. *IEEE Trans. Antennas Propag.* **2018**, *66*, 7412–7417. [CrossRef]
16. Sun, L.; Feng, H.; Li, Y.; Zhang, Z. Compact 5G MIMO Mobile Phone Antennas with Tightly Arranged Orthogonal-Mode Pairs. *IEEE Trans. Antennas Propag.* **2019**, *66*, 6364–6369. [CrossRef]
17. Ren, Z.; Zhao, A.; Wu, S. MIMO Antenna with Compact Decoupled Antenna Pairs for 5G Mobile Terminals. *IEEE Antennas Wirel. Propag. Lett.* **2019**, *18*, 1367–1371. [CrossRef]
18. Moses, A.; Moses, N.; Janapala, D. An Electrically Small 4-Port Self-Decoupled MIMO Antenna Pairs Operating in n78 5G NR Band for Smartphone Applications. *AEU Int. J. Electron. Commun.* **2022**, *145*, 154082.
19. Huang, J.; Chen, Z.; Cai, Q.; Loh, T.H.; Liu, G. Minimized Triple-Band Eight-Element Antenna Array for 5G Metal-frame Smartphone Applications. *Micromachines* **2022**, *13*, 136. [CrossRef]
20. Chandel, R.; Gautam, A.K.; Rambabu, K. Design and Packaging of an Eye-Shaped Multiple-Input–Multiple-Output Antenna with High Isolation for Wireless UWB Applications. *IEEE Trans. Comp. Pack. Man. Technol.* **2018**, *8*, 635–642. [CrossRef]
21. Nandiwardhana, S.; Chung, J. Trade-Off Analysis of Mutual Coupling Effect on MIMO Antenna Multiplexing Efficiency in Three-Dimensional Space. *IEEE Access* **2018**, *6*, 47092–47101. [CrossRef]

22. Zhao, A.; Ren, Z. Size Reduction of Self-Isolated MIMO Antenna System for 5G Mobile Phone Applications. *IEEE Antennas Wirel. Propag. Lett.* **2019**, *18*, 152–156. [CrossRef]
23. Zhang, X.; Li, Y.; Wang, W.; Shen, W. Ultra-Wideband 8-Port MIMO Antenna Array for 5G Metal-Frame Smartphones. *IEEE Access* **2019**, *7*, 72273–72282. [CrossRef]
24. Cheng, B.; Du, Z. A Wideband Low-Profile Microstrip MIMO Antenna for 5G Mobile Phones. *IEEE Trans. Antennas Propag.* **2022**, *70*, 1476–1481. [CrossRef]
25. Li, R.; Mo, Z.; Sun, H.; Sun, X.; Du, G. A Low-Profile and High-isolated MIMO Antenna for 5G Mobile Terminal. *Micromachines* **2020**, *11*, 360. [CrossRef]

Article

Design of a Differential Low-Noise Amplifier Using the JFET IF3602 to Improve TEM Receiver

Shengjie Wang [1,2], Yuqi Zhao [1,2], Yishu Sun [1,2], Weicheng Wang [1,2], Jian Chen [1,2] and Yang Zhang [1,2,3,*]

1. Key Laboratory for Geophysical Instrument of Ministry of Education, Jilin University, Changchun 130061, China
2. College of instrumentation & Electrical Engineering, Jilin University, Changchun 130061, China
3. Engineering Research Center of Geothermal Resources Development Technology and Equipment, Ministry of Education, Jilin University, Changchun 130026, China
* Correspondence: zhangyang19@jlu.edu.cn

Abstract: The observed data of transient electromagnetic (TEM) systems is often contaminated by various noises. Even after stacking averages or applying various denoising algorithms, the interference of the system noise floor cannot be eliminated fundamentally, which limits the survey capability and detection efficiency of TEM. To improve the noise performance of the TEM receiver, we have designed a low-noise amplifier using the current source long-tail differential structure and JFET IF3602 through analyzing the power spectrum characteristics of the TEM forward response. By the designed circuit structure, the JFET operating point is easy to set up. The adverse effect on the JFET differential structure by JFET performance differences is also weakened. After establishing the noise model and optimizing the parameters, the designed low-noise differential amplifier has a noise level of $0.60 \text{nV}/\sqrt{\text{Hz}}$, which increases the number of effective data 2.6 times compared with the LT1028 amplifier.

Keywords: low noise; junction field-effect transistor; transient electromagnetic method

1. Introduction

The transient electromagnetic (TEM) method is a classic geophysical exploration method that detects the induction response of a geological body to emitted electromagnetic (EM) waves. It distinguishes subsurface geological structures based on characteristic differences in the amplitude and decay rate of the response of geological bodies with different resistivities. Due to the non-destructive propagation of EM waves in the subsurface medium, TEM is widely used in mineral resource exploration, geological surveys, and urban disease exploration. The TEM signal is characterized by high amplitude in the early time and low amplitude in the late time. Therefore, in order to obtain a high-quality signal, it is necessary to use a wide frequency band, a large dynamic range amplifier, and a high-precision acquisition system with high speed to collect high-quality TEM data for subsequent data processing and data interpretation [1,2].

The observed signal is always interfered with by strong noise in practice. The noise contaminates the signal, especially the late time, which is in low amplitude and represents the deep geological information [3]. These kinds of noise are classified by their source. Part of the noise is from the EM interference in the environment, which is called environmental noise; the other part of the noises is from the TEM receiver, which is called the system noise floor. Environmental noise has a certain pattern in statistics and can be eliminated by using algorithms such as half-cone gate filtering, minimum noise separation, and improved algorithms based on temporal correlation [4–6]. However, these data processing methods cannot eliminate the effect of the system noise floor, so the TEM receiver needs to be optimized to obtain a more accurate signal.

The amplifier noise floor of the TEM receiver is a major part of the system noise floor, and there have been many related studies on TEM low-noise amplifiers (LNA) [7]. Chen et al. designed an amplifier with a noise floor of $1.83\text{nV}/\sqrt{\text{Hz}}$ using the low-noise integrated operational amplifier (IOA) AD797 for the ZTEM receiving device [8]. Pi et al. used the IOA LT1028 to design a low-noise preamplifier circuit with an ideal noise of $2.45\text{nV}/\sqrt{\text{Hz}}$ for urban TEM devices [9]. Low-noise IOA is mainly used in TEM preamplifiers currently [10,11], but the noise characteristics of IOAs are inferior to discrete components due to the cost restriction, which also limits the noise optimization of subsequent circuits [12]. To solve the problem, Wang et al. developed a $2\text{nV}/\sqrt{\text{Hz}}$ ZTEM signal conditioning circuit using the junction field-effect transistor (JFET) LSK389B in the helicopter TEM receiver, which introduced a new idea for the optimization of TEM devices [13].

At present, JFET LNA is mainly used for the measurement and amplification of weak signal sensors such as piezoelectric accelerometers, wireless radio frequency equipment, and seismic accelerometers at present [14]. Scandurra proposed a feedback compensation JFET differential amplifier circuit for low-frequency noise measurement circuits with the noise floor of $1.00\text{nV}/\sqrt{\text{Hz}}@1.00\text{kHz}$ [15]. Cannatà designed a $0.80\text{nV}/\sqrt{\text{Hz}}@1.00\text{kHz}$ single-ended amplifier based on discrete components JFET for low-frequency noise measurement [16]. To summarize, JFET has a very low system noise floor and a wide stable noise frequency band [17], which is very suitable for noise optimization of TEM receivers.

The rest of the paper is as follows: In Section 2, we analyze the time-frequency characteristics of the TEM signal according to the TEM principle and forward response, and clarify the requirements; Section 3 gives the JFET LNA for TEM receiver with the circuit structure, model, and actual noise floor test; In Section 4, we conducted laboratory experiments to compare the length of TEM effective data from the JFET receiver and the IOA receiver. Finally, the optimization direction of the amplifier and the application effect in transient electromagnetic systems are discussed.

2. The Analysis of TEM Signal

The TEM system consists of a transmitting coil, a transmitter, a receiving coil, and a receiver. Its working principle is shown in Figure 1. A bipolar step pulse current is sent through the transmitting coil by the transmitter. Meanwhile, a primary field is created, surrounded, and transmitted in the form of smoke rings. Due to Faraday's law of electromagnetic induction, geological bodies are excited by the primary field, and the eddy current is induced underground with the secondary field when the pulse is on. When the pulse is turned off, the secondary field starts decaying at different rates, which are related to the conductivity of the subsurface layers. The voltage in the receiving coil is generated by the secondary field and observed in the receiver. After TEM data interpretation, the abnormality and subsurface layers can be known [18].

To further analyze the characteristics of TEM signals, we use the TEM forward modeling to generate ideal signals for time-frequency analysis. Since the actual geological layering is very complex, it is difficult to fully cover the condition of multi-layer forward modeling. Because the TEM has the characteristics of a strong response in a low-resistance medium and a weak response in a high-resistance medium, the actual geological response will be higher than the pure high-resistance response and lower than the pure low-resistance response. Therefore, we set the high-resistance uniform half-space to $100\ \Omega\cdot\text{m}$ and adjust the transmitting magnetic moment to $1 \times 10^4\ \text{A}\cdot\text{m}^2$, and the receiving equivalent area to $128\ \text{m}^2$ [19]. The forward modeling based on the sinusoidal numerical filtering algorithm is used to obtain the forward response within 1 ms, as shown in Figure 2 [20,21]. It is observed that the early TEM forward response is five orders of magnitude higher than the late response. Therefore, to identify the late signal, not only a high-amplification amplifier is required, but also the noise floor of the amplifier is limited; otherwise, it is extremely difficult to obtain useful information on TEM from the circuit noise. Since the noise power spectrum is mainly used to measure the noise characteristics of LNA, we calculate the power spectral density (PSD) of the forward response in $100\ \Omega\cdot\text{m}$ pure high resistance as a reference index for designing the LNA. The results in Figure 3 show that the low-frequency PSD of the TEM signal is larger and the high-frequency PSD is smaller and below $1nV/\sqrt{Hz}$. Therefore, when designing the amplifier, it is necessary to ensure that the amplifier has a stable gain and an extremely low noise floor in the whole frequency band to obtain high-quality TEM data and lay the foundation for subsequent data denoising and data interpretation.

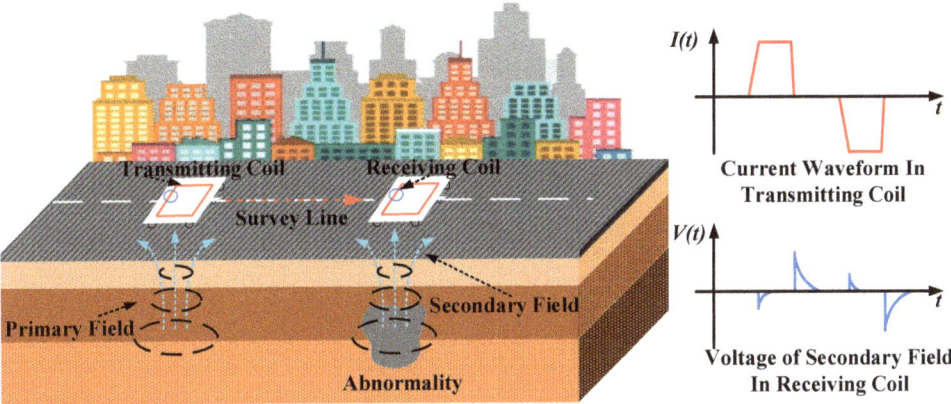

Figure 1. Schematic diagram of TEM survey.

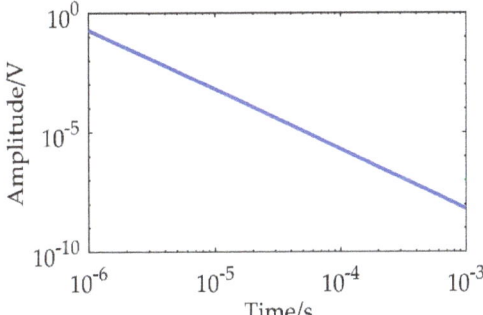

Figure 2. The TEM forward response.

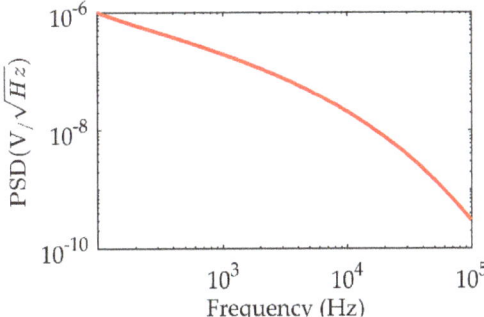

Figure 3. The PSD of TEM forward response.

3. Optimal Design of JFET LNA

3.1. Selection of LNA Circuit Components

Because of the characteristics of the wideband and large dynamic range of the TEM signal, the suitable amplifier must have a sufficient gain-bandwidth product, as mentioned, and a low-noise cascade amplifier circuit needs to be designed to achieve reliable signal amplification. The Frith formula, shown in Equation (1), reflects the relationship between the overall noise of the cascaded circuit and the noise of the circuits at all levels as follows:

$$F = F_1 + \frac{F_2 - 1}{K_1} + \frac{F_3 - 1}{K_1 K_2} + \cdots + \frac{F_M - 1}{K_1 K_2 \cdots K_{M-1}} \quad (M = 2, 3, \cdots) \quad (1)$$

F is the total noise figure of the cascaded amplifier, F_M is the noise figure of the M amplifier, and K_{M-1} is the gain of the $M-1$ amplifier. The noise of the cascaded amplifier is mainly restricted by the first-stage amplifier, and the influence of the latter stage is slighter, which means reducing the noise floor of the first-stage amplifier is the primary consideration when designing a low-noise system for weak signal detection.

To optimize the noise floor of the first-stage amplifier circuit, a low-noise device must be selected as the core of the circuit. Common choices of low-noise devices are bipolar transistors (BJTs), JFETs, and IOAs. Because BJT devices and BJT-type IOAs have a low noise floor in low-frequency and mid-frequency bands, they are generally used to make TEM preamplifiers, but with the development of low-noise JFET, such as LSK389B and 2SK3320. These components significantly reduce device voltage noise when their own current noise is negligible. These give low-noise JFETs the advantage of a low device noise floor and simple noise components, which are more suitable for making LNAs. In order to select the low-noise device for the preamplifier, we compared the noise performance of common low-noise devices, and the results are shown in Table 1.

Table 1. Comparison of noise performance of low-noise devices.

Device Model	e_n (nV/$\sqrt{Hz}$)	i_n (pA/$\sqrt{Hz}$)
AD745	3.2 (f = 1 kHz)	0.007 (f = 1 kHz)
AD797	0.9 (f = 1 kHz)	2.0 (f = 1 kHz)
AD8672	2.8 (f = 1 kHz)	0.3 (f = 1 kHz)
OPA847	0.92 (f > 1 MHz)	3.5 (f > 1 MHz)
INA163	1.0 (f = 1 kHz)	0.8 (f = 1 kHz)
LT1028	0.85 (f = 1 kHz)	1.0 (f = 1 kHz)
LSK389	0.9 (f = 1 kHz)	—
IF3602	0.5 (f = 100 Hz)	—

Table 1 shows that IF3602 has very low voltage noise (0.5nV/$\sqrt{Hz}$@100Hz) and the noise frequency is only 100 Hz. The principle of a JFET is using the electric field effect in the semiconductor to change the barrier width of the gate PN junction by controlling the gate-source voltage V_{GS} and using the pinch-off of the depletion layer to control the internal carriers. No current flows between the gate and the source, so the current noise is negligible. This feature simplifies the noise model of the JFET amplifier circuit. According to the comparison, we proposed a low-noise differential circuit design for the JFET IF3602 and combined the noise model to optimize the external circuit.

3.2. Overall Design of Low-Noise JFET Amplifier Circuit

Based on the characteristics of the TEM signal analyzed in Figures 2 and 3, we designed the amplifier circuit shown in Figure 4. According to the functions of each cascade amplifier, the LNA circuit is divided into a four-level structure, as shown in Figure 5. The first-stage circuit is a differential circuit based on the JFET IF3602, which can achieve low-noise and high-gain amplification for the received signal. The low noise floor of the amplifier circuit means that the impact of circuit noise on the signal-to-noise ratio (SNR) of the TEM signal is weak. Besides, the high gain in the first-stage circuit weakens the impact of the noise floor in the post-stage circuits. For the differential circuit, it also has the characteristics of high differential mode magnification and low common mode magnification, which can effectively suppress the interference of common mode noise and enhance the anti-interference ability of the TEM device.

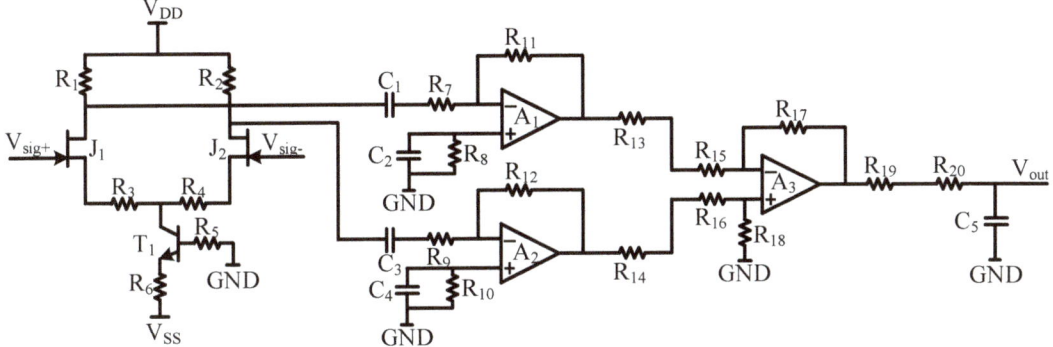

Figure 4. Schematic diagram of low-noise amplifier circuit.

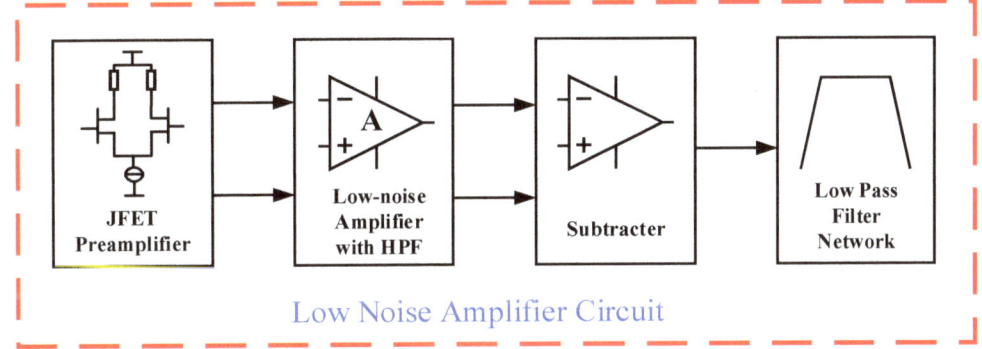

Figure 5. Schematic diagram of the function of the low-noise amplifier circuit.

The second-stage circuit is an amplifier made from the low-noise IOA LT1028 with a high-pass filtering function. Although the noise floor of the LT1028 IOA is not as good as IF3602, its equivalent voltage noise is only 0.85nV/$\sqrt{Hz}$, and the IOA circuit is easy to design because of its simple structure. Therefore, it is very suitable to be the secondary amplifier. In addition, as shown in Figure 3, it is necessary to reduce the impact of a low-frequency component of TEM to avoid the early signal saturation. Therefore, adding high-pass filtering can make full use of its amplitude-frequency characteristics, to reduce the amplification factor of low-frequency components, and avoid early signal saturation.

The third-stage circuit contains a precision differential amplifier INA105, which integrates a laser-corrected 25 kΩ precision resistor to ensure the gain accuracy of the internal feedback amplifier circuit and reduce the impact of resistance temperature drift on the accuracy of the differential amplifier.

An RC low-pass filter network is used in the fourth stage with a 257 kHz cut-off frequency, which includes the ideal signal frequency band in forwarding modeling, to avoid the influence of high-frequency noise on the circuit and ensure the quality of TEM signals.

3.3. Design and Analysis of JFET Differential Low-Noise Amplifier Circuit

As an important part of the LNA, the JFET differential amplifier circuit needs to have the characteristics of low noise and high gain. As a voltage-controlled device, JFET must work in the amplification mode when used for signal amplification. At that time, the depletion layer is partially pinch-off, the input AC signal causes the width of the depletion layer to change slightly, and the output drain current also changes accordingly. The signal is amplified at the output. The noise floor and amplification capability of JFET are also limited by the internal channel. The JFET must be operated at a suitable static operating point to make it both low-noise and high-gain.

In practice, due to the limitations of manufacturing technology and semiconductor doping technology, JFET devices will show a large performance difference even in the same batch [17], resulting in the unequal current in the left and right branches of traditional JFET long-tail differential circuits. This problem makes the circuit design difficult because of the static operating point offset. Fortunately, the IF3602 not only has excellent amplification characteristics and noise performance but also integrates a pair of matched low-noise JFETs, which greatly reduces the impact of JFETs differences on performance. In addition, we introduce a source-coupled differential amplifier circuit design, in which a current source is used to replace the tail resistors in traditional long-tail differential circuits. Compared with the long-tail differential circuit, the higher equivalent resistance of the current source gives the circuit an extremely strong common-mode signal rejection and can provide stable current output for the branch. Usually, the following three correlated variables need to be considered when setting the static operating point of the JFET: drain-source voltage V_{DS}, gate-source voltage V_{GS}, and drain current I_D. However, the new source-coupled differential amplifier circuit design reduces the variables to V_{DS} and V_{GS}, which simplifies the difficulty of circuit design. The DC path and the AC path were analyzed in this part, where the DC path is used to stabilize the static operating point for the device for optimum performance of the JFET; the AC path determines the amplification capability of the circuit. The results will provide theoretical support for subsequent circuit noise analysis and circuit optimization.

The DC path of the JFET amplifier circuit is shown in Figure 6, where BJT T_1, R_5, and R_6 are formed as a current source to output a stable current of $I_p = (-V_{SS} - V_{PN})/R_6$. Due to the high symmetry of the circuit, $I_{D1} = I_{D2} = 0.5 I_p$. After determining the quiescent operating current of the JFET, only the operating voltages V_{DS} and V_{GS} need to be considered. For JFET, the stable amplification must be achieved in the amplification mode. It is necessary to keep the $V_{DS} > 0$, $V_{GS} < 0$ where $V_D = V_{DD} - I_{D1}R_1 = V_{DD} - I_{D2}R_2$ and $I_{D1}R_3 = -V_{BE} + I_{D2}R_4$. The JFET gate is grounded during DC analysis, so $V_G = 0$. The current source maintains the stability of the loop current, so I_{D1} and I_{D2} are known. According to the external resistors R_1, R_2, R_3, and R_4, the static operating point of the amplifier circuit can be calculated, avoiding the problem of static operating point setting caused by the interaction of V_{GS}, V_{DS}, and I_D in the JFET circuit. In order to ensure that the IF3602 has excellent noise characteristics, it needs to be set at the minimum noise static operating point "$V_{DS} = 3$ V, $I_D = 5$ mA" in the datasheet. After calculating the theoretical external parameters, the R_1 and R_2 should be fine-tuned to fit the minimum noise static operating point in the datasheet as much as possible, ensuring that the IF3602 can stably exert its noise performance advantages.

The AC path of the JFET amplifier circuit is shown in Figure 7. Because the differential circuit contains excellent differential mode signal amplification capability and common mode noise suppression capability, the current source at the common terminal does not be amplified. Therefore, the current source can be ignored in the AC equivalent circuit. Besides, the amplification capabilities of the circuits on both sides are the same, so only one side of the circuit needs to be considered when constructing the small-signal model to analyze the magnification. The single-side small-signal model is shown in Figure 8. It can be seen from Figure 8 that the amplification factor of the differential amplifier circuit satisfies the following:

$$A_v = (-g_m R_1)/(1 + g_m R_3) = (-g_m R_2)/(1 + g_m R_4) \qquad (2)$$

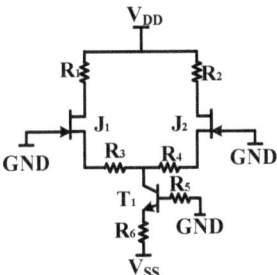

Figure 6. DC equivalent path of JFET differential circuit.

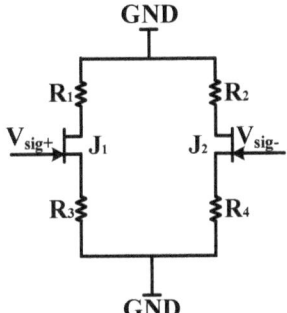

Figure 7. AC equivalent path of JFET differential circuit.

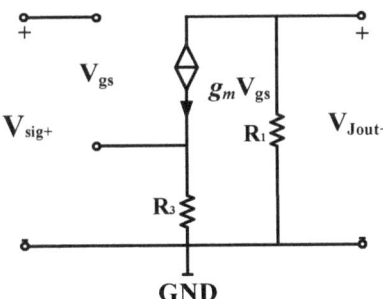

Figure 8. Analysis of AC small signal model of JFET differential circuit.

g_m is the JFET transconductance, which is affected by the internal conduction channel, so it is only related to the static operating point. The source-coupled differential amplifier circuit constructed by the current source not only simplifies the circuit design but also stabilizes the noise performance and amplification capability of the JFET IF3602. A_v is the magnification of the differential amplifier circuit, and its value mainly depends on the ratio of the external circuits R_1 and R_3 or R_2 and R_4. The purpose of R_3 and R_4 is to stabilize the source potential of the JFET. In order to maintain the amplification ability of the circuit, their values are relatively small (less than 10 Ω). Therefore, maintaining the JFET in the amplification area, keeping the JFET working near the static operating point of minimum noise, and increasing the values of R_1 and R_2 at the same time will significantly increase the amplification factor of the differential amplifier circuit and ensure the performance of the first-stage differential amplifier circuit.

3.4. Noise Analysis and Parameter Optimization of JFET Differential Amplifier Circuit

The equivalent model of the circuit is constructed by analyzing the AC and DC path signals of the circuit. In order to further optimize the noise performance and amplification performance of the JFET IF3602 differential amplifier circuit, a noise equivalent model was built as shown in Figure 9. e_{n1} and e_{n2} are the equivalent noise sources of the IF3602, e_{n3} and e_{n4} are the equivalent thermal noise sources of the drain resistors R_1 and R_2, and e_{n5} and e_{n6} are the equivalent thermal noise sources of the source resistors R_3 and R_4. T_1, R_5, and R_6 constitute the current source I. According to the previous analysis of the AC path, the current source does not get amplified as the AC signal, and its equivalent noise source is suppressed by the differential amplifier circuit as a common-mode signal, which can be ignored.

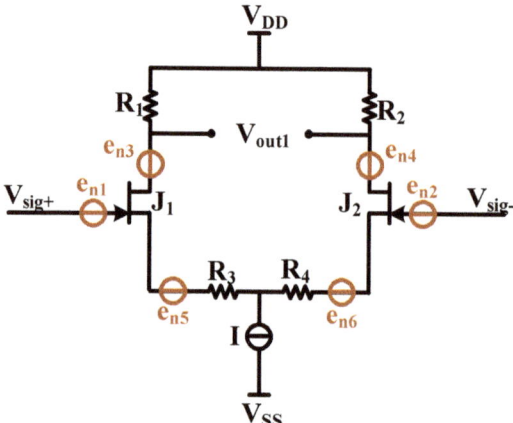

Figure 9. JFET differential circuit noise model.

For the equivalent noise sources e_{n1} and e_{n2} of IF3602, the noise source is input from the gate, and the differential amplifier circuit can be equivalent to a common source amplifier circuit, and its amplification factor is $(-g_m R_2)/(1+g_m R_4)$. The sum of the output noise of e_{n1} and e_{n2} at the output of V_{out1} is as follows:

$$\sqrt{2(-g_m R_1)^2 e_{n1}^2/(1+g_m R_3)^2} \tag{3}$$

For thermal noise sources e_{n3} and e_{n4} of R_1 and R_2, the noise source PSD satisfies $\sqrt{4kTR_1}$ and $\sqrt{4kTR_2}$, k is the Boltzmann constant, and T is the temperature in Kelvin. These sources influence on the output end directly, and R_1 and R_2 are equal, so the sum of the output noise of e_{n3} and e_{n4} at the output end is as follows:

$$\sqrt{2e_{n3}^2} \tag{4}$$

The PSDs of thermal noise sources e_{n5} and e_{n6} of R_3 and R_4 are $\sqrt{4kTR_3}$ and $\sqrt{4kTR_4}$, respectively, and their resistance values are equal. The thermal noise is input from the source of IF3602 and output at the drain after being amplified. In the circumstances, the differential circuit should be equivalent to a common-gate amplifying circuit, and its amplification factor is $g_m R_1/(1+g_m R_3)$. The sum of the output noises of e_{n5} and e_{n6} at the output is as follows:

$$\sqrt{2(g_m R_1)^2 e_{n5}^2/(1+g_m R_3)^2} \qquad (5)$$

According to the circuit superposition theorem, the sum of the output noise of the IF3602 differential amplifier circuit at the output end is as follows:

$$e_{nJFETout} = \sqrt{2e_{n3}^2 + 2(e_{n1}^2 + e_{n5}^2)(g_m R_1)^2/(1+g_m R_3)^2} \qquad (6)$$

Usually, to avoid the influence of circuit amplification on noise results, it is necessary to normalize the output noise by dividing it by the circuit's amplification (i.e., the transfer function) and convert it into equivalent input noise to measure the noise floor of the circuit. The equivalent input noise of the differential amplifier circuit is as follows:

$$e_{nJFETin} = \frac{e_{nout}}{|A_v|} = \sqrt{2e_{n1}^2 + 2e_{n5}^2 + 2e_{n3}^2(1+g_m R_3)^2/(g_m R_1)^2} \qquad (7)$$

The equivalent noise source e_{n1} of the JFET is related to the static operating point, which can be regarded as a fixed value after referring to the minimum noise static operating point setting in the datasheet. The value of R_3 is very small and leads to the thermal noise introduced into the input is extremely limited. Therefore, the optimization of resistor R_1 should be focused. From Equation (6), the thermal noise of R_1 resistance at the input terminal equivalently is as follows:

$$e_{nR1in} = \sqrt{e_{n3}^2(1+g_m R_3)^2/(g_m R_1)^2} = \sqrt{4kT(1+g_m R_3)^2/g_m^2 R_1} \qquad (8)$$

When increasing the resistance value of R_1, the input noise of R_1 will decrease according to Equation (8), and the amplification factor will increase according to Equation (2). The whole optimization meets the requirements of the increasing the gain of the first stage and decreasing the noise floor of the circuit. However, increasing R_1 will inevitably cause the offset of the static operating point of the circuit; therefore, the supply voltage of the amplifier circuit must be increased synchronously to roughly keep the static operating point of the JFET as described in the datasheet (V_{DS} = 3V, I_D = 5mA). After optimization, the parameters of the amplifier circuit shown in Figure 4 are finally determined as shown in Table 2.

Table 2. Circuit device parameters.

Components	Component Type	Component Value
R_1, R_2	1% Resistor 2.4 kΩ 0805 × 2	1.2 kΩ
J_1, J_2	IF3602 Low Noise N-Channel JFET Pair	InterFET
R_3, R_4	1% Resistor 0805	1 Ω
T_1	BJT	2N1711
R_5	1% Resistor 0805	0 Ω
R_6	1% Resistor 3.6 kΩ 0805 × 4	900 Ω
C_1, C_3	4.7 µF Y5V 0805 Capacitor × 4	18.8 µF
C_2, C_4	Y5V 0805 Capacitor	10 nF
R_7, R_9	1% Resistor 0805	510 Ω
R_8, R_{10}	1% Resistor 0805	100 Ω
A_1, A_2	Low Noise IOA	LT1028
R_{11}, R_{12}	1% Resistor 0805	5.1 kΩ
R_{13}, R_{14}, R_{19}	1% Resistor 100 Ω 0805 × 2	50 Ω
R_{15}, R_{16}, R_{17}, R_{18}	INA105 Built-in Resistor	25 kΩ
A_3	Precision Difference Amplifier	INA105
R_{20}	1% Resistor 0805	910 Ω
C_5	Y5V 0805 Capacitor	860 pF
V_{DD}	Lithium Battery	+12 V
V_{SS}	Lithium Battery	−12 V

3.5. Noise Testing and Analysis

The instrument can only measure the output noise power spectrum of the amplifier circuit. The actual noise floor of the sensing system needs to be converted through the transfer function. First, the transfer function of each level, the overall transfer function in Figure 4, needs to be calculated, and the results are shown in Equations (9)–(13) as follows:

$$H_{circuit}(s) = \frac{V_{out}(s)}{V_{in}(s)} = \frac{V_{out}(s)}{V_{sig+}(s) - V_{sig-}(s)} = H_1(s)H_2(s)H_3(s)H_4(s) \quad (9)$$

$$H_1(s) = A_v = (-g_m R_1)/(1 + g_m R_3) = (-g_m R_2)/(1 + g_m R_4) \quad (10)$$

$$H_2(s) = (-sC_1 R_{11})/(1 + sC_1 R_7) = (-sC_3 R_{12})/(1 + sC_3 R_9) \quad (11)$$

$$H_3(s) = R_{17}/(R_{13} + R_{15}) = R_{18}/(R_{14} + R_{16}) \quad (12)$$

$$H_4(s) = 1/(1 + sC_5(R_{19} + R_{20})) \quad (13)$$

The final transfer function can be calculated by substituting the parameters in Table 2 into Equation (9). The result is shown in Figure 10.

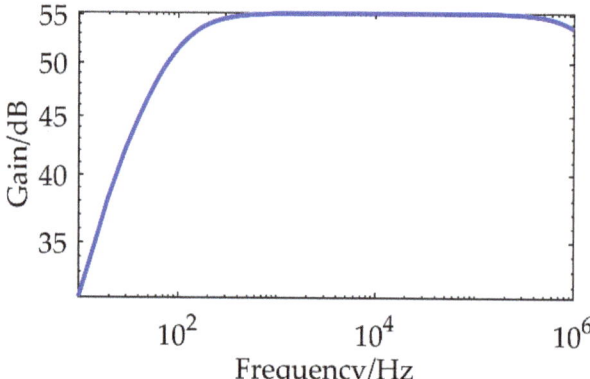

Figure 10. Overall amplitude-frequency characteristic curve of low-noise amplifier circuit.

The gain of the IF3602 LNA circuit is kept stable at 55 dB in the 100 Hz–100 kHz. Because the PSD of the TEM low-frequency signal is large, if the gain at the full band is maintained stable, the signal will be more likely to saturate in the early stage. In this case, a non-ideal high-pass filter can be used to reduce the gain of the amplifier circuit for low-frequency signals, because the gain of the non-ideal high-pass filter is weak in the low frequency and large in the high frequency. A more accurate TEM secondary field signal can be recovered by adding a transfer function correction during data processing. The noise PSD at the output of the circuit was detected using an Agilent 35670A dynamic signal analyzer when shorted the input under EM shielding conditions. In order to record the noise floor of the amplifier circuit detail, the output noise floor of the IF3602 low-noise amplifier and LT1028 amplifier in the frequency bands of 0–800 Hz, 800–1600 Hz, and 1.6–52.8 kHz were collected and spliced. The equivalent input noise floor of the two is shown in Figure 11.

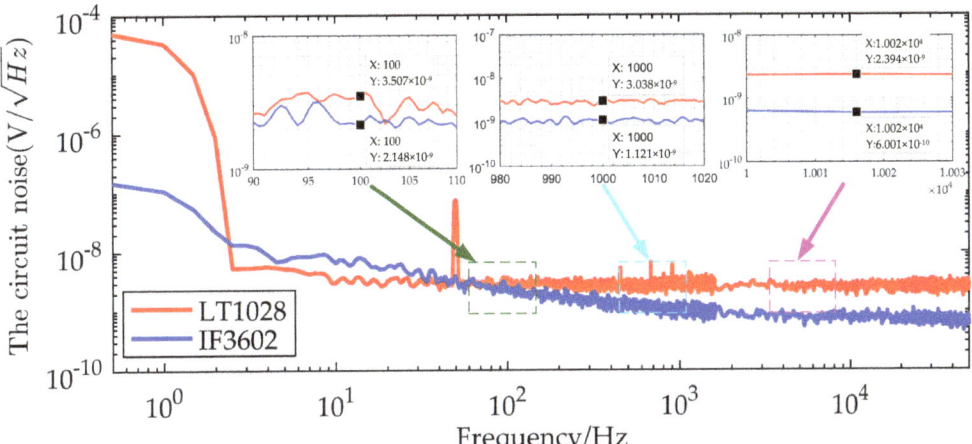

Figure 11. Comparison of the noise floor between the IF3602 circuit and the LT1028.

According to Figure 11, the noise floor of the IF3602 differential amplifier at 100 Hz, 1 kHz, and 10 kHz are 2.15nV/$\sqrt{Hz}$, 1.12nV/$\sqrt{Hz}$, and 0.60nV/$\sqrt{Hz}$ respectively, which are lower than the corresponding frequency noise of the LT1028 as follows: 3.50nV/$\sqrt{Hz}$, 3.04nV/$\sqrt{Hz}$, and 2.39nV/$\sqrt{Hz}$. The noise floor of the IF3602 circuit is significantly lower than the noise floor of the LT1028 circuit in the TEM band of 100 Hz–52.8 kHz.

Figure 12 compares the actual noise curve of the IF3602 differential amplifier circuit with the calculated one based on Table 2 and Equation (7). The actual noise is close to the theoretical noise, and both of them reach below 1nV/$\sqrt{Hz}$ above 1 kHz. However, the noise floor is different at low frequencies. The reason is that, when calculating the thermotical noise, only the corresponding noise of the device at 100 Hz given by the datasheet was used. However, the JFET is seriously disturbed by $1/f$ noise at low frequency in practice, and the amplitude of $1/f$ noise is inversely proportional to the frequency. So, in the high-frequency band, the theoretical noise is close to the actual noise, and the actual noise floor deviates from the theory.

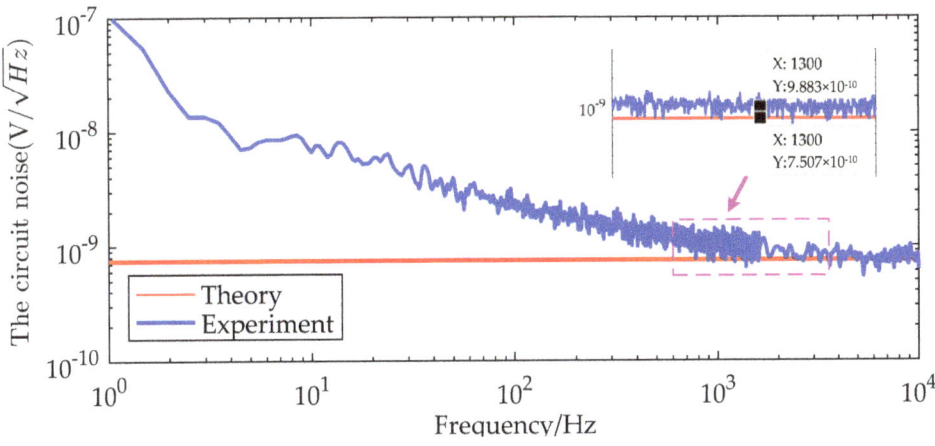

Figure 12. Comparison of measured noise data and theoretical calculated values of IF3602 low-noise amplifier circuit.

4. Indoor Comparison Experiment

To further compare the performance of the IF3602 LNA in the receiver, we experimented with the simulation using a small-loop TEM device in the laboratory. Figure 13 shows the schematic diagram of the comparison test. The devices are shown in Figure 14, and the parameters of the device are as follows:

- The radius of the small loop transmitting coil used is 26.5 cm;
- The number of turns of the transmitting loop is 10;
- The transmitting current is 6 A;
- The radius of the receiving coil is 10 cm;
- The number of turns of the receiving loop is 32.

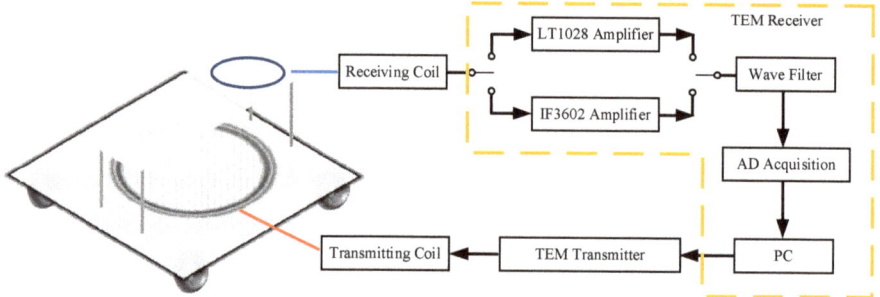

Figure 13. Schematic diagram of laboratory simulation comparison test.

To keep the circuit filter from interfering with data observation, a low-noise LT1028 amplifier board with an amplification factor of 10 was used as the pre-stage, and the same post-stage circuit as the IF3602 low-noise circuit was added, which was adjusted to have the same frequency band and similar gain as the IF3602 amplifier. Both groups of experiments were taken 128 times superposition average to avoid the influence of the amount of superposition on the quality of the received data. The results are shown in Figure 15.

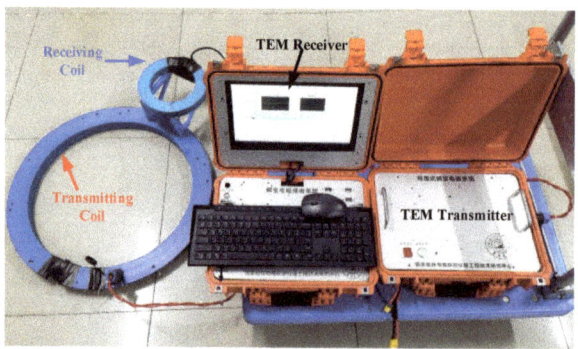

Figure 14. The laboratory simulation comparison test and TEM devices.

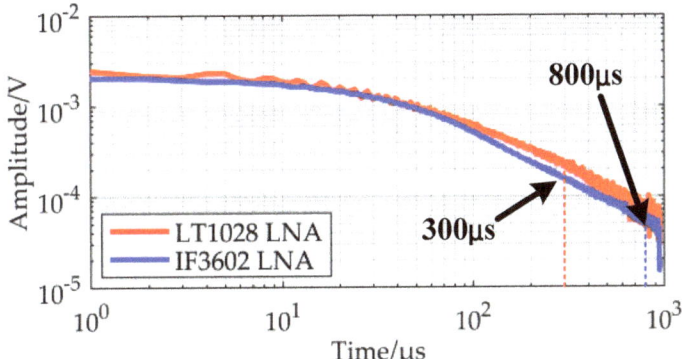

Figure 15. Comparison of the data observed by the receiving device in the laboratory.

According to the results, the IF3602 amplifier receiver has a smoother signal and higher data quality than the LT1028 amplifier receiver after taking a double logarithm. The effective data length of the IF3602 amplifier raw data is 800 µs, which is about 2.6 times longer than the LT1028 amplifier receiver.

The relationship between TEM detecting depth and detecting time is described using formula $h = \sqrt{2t/\sigma\mu_0}$, σ is the electrical conductivity of the geological body and $\mu 0$ is the vacuum magnetic permeability. Since the location of the experiment remains unchanged, the electrical conductivity values of the two geological bodies are the same, and the vacuum magnetic permeability is a fixed value, so the performance of the system is only determined by the effective length of the data. The receiver with the IF3602 amplifier increases the TEM theoretical maximum detection depth to about 1.6 times that of the original system, according to the result and the TEM detecting depth formula. In addition, because of the smoother performance of the IF3602 receiver observed data under 128 times of superposition, it is suggested that the number of superpositions can be reduced to have the same SNR as the LT1028 amplifier receiver while still accessing the higher detection efficiency.

5. Discussion

The IF3602 LNA has better noise characteristics than the current TEM common LT1028 LNA, and its noise reduces to 0.60nV/$\sqrt{Hz}$ at 10 kHz. By comparing the IF3602 LNA TEM receiver and the LT1028 LNA TEM receiver, the observed signal of the receiver with the IF3602 LNA has a significant improvement in effective data length and data quality. The detection depth is increased to 1.6 times that of the former LT1028 system. However, we hold that there are still the following problems, which limit its application and use effect.

Due to the limitations of the JFET device manufacture, the parameters between the devices are quite different, which requires strict pairing before use. Even if the IF3602 used in the experiment has been paired, there are still differences in the internal JFET. The symmetry of the circuit is not as good as that of the ideal differential circuit. The ignored current source also introduces circuit noise. Although the IF3602 has a pair of paired JFETs built-in, its high cost and its inside JFETs parameter difference limit the application of low-noise JFETs.

However, we hold that the above problems can be ameliorated by the following improvements. Before the circuit is manufactured, it is recommended to use a graphic instrument to test the JFET parameter curve and try to select a JFET with similar parameters. If the test conditions are not available, i.e., the parameter characteristic curve cannot be tested, first, a simple circuit can be designed about the minimum noise static operating point in the datasheet, then measure the actual static operating point of each JFET, and select a JFET with good consistency. This can effectively reduce the introduction of common-mode circuit noise caused by JFET parameter difference. Matching can effectively reduce the demand for JFETs such as the IF3602 that have been matched by manufacturers and can effectively reduce the cost of JFET LNA circuits.

6. Conclusions

In order to improve the survey capability and accuracy of the TEM system, we have designed a low-noise differential amplifier for the TEM receiver by using the low-noise JFET IF3602 according to the power spectrum characteristics of the TEM forward response. Additionally, effectively increase the effective length and quality of the TEM observed data.

Firstly, based on the model of TEM forward response, we created the theoretical signal by homogeneous half-space. Secondly, we analyzed the power spectrum characteristics of the TEM signal to determine the requirements of the TEM amplifier noise floor and frequency bands. Thirdly, we designed a source-coupled differential amplifier circuit using the IF3602 low-noise JFET. We applied a current source to replace the resistance in the conventional long-tail circuit for providing stable current output for the left and right branches to simplify the JFET static working point settings. Besides, the parameters of the low-noise circuit were optimized combined with the theoretical model of the circuit and noise. Results of the noise test indicate that the designed circuit achieves good noise characteristics performance with a minimum of $0.60 \text{nV}/\sqrt{\text{Hz}} \cdots$ @10kHz, which is significantly lower than the performance of the corresponding frequency point of the LT1028 LNA. It can improve the ability of the system to distinguish TEM signals. Compared with the LT1028 LNA, the designed IF3602 low-noise circuit can increase the effective data points of the observed data by about 2.6 times and the theoretical detection depth by about 1.6 times under the same experimental conditions. In addition, under the condition of the same number of superpositions, the signal amplified by the IF3602 LNA is of high quality and smoother, which proves that reducing the noise floor of the preamplifier can not only improve the data quality but also greatly help improve the survey capability and detection efficiency.

Therefore, this study is of great significance for the design of preamplifier circuits in TEM receivers. In future work, we will continue to explore new circuit design schemes, such as parallel noise reduction design, to further improve the detection performance of the TEM receivers.

Author Contributions: S.W.: JFET amplifier design, experiment design, writing, discussion. Y.Z. (Yuqi Zhao): noise floor test, LNA circuit making, discussion. Y.S.: the PSD programing, discussion, English editing. W.W.: discussion, experiment assistance, English editing. J.C.: discussion, TEM data processing. Y.Z. (Yang Zhang): research guidance, editing, discussion. All authors have read and agreed to the published version of the manuscript.

Funding: This research was supported by the National Natural Science Foundation of China (Grant Nos. 41827803), Jilin University Youth Teacher Student Cross disciplinary Cultivation Project (2022-JCXK-32) and the Engineering Research Center of Geothermal Resources Development Technology and Equipment, Ministry of Education, Jilin University.

Data Availability Statement: Upon reasonable request, the data supporting this investigation are available from the corresponding authors.

Conflicts of Interest: The authors declare no conflict of interest.

References

1. Chen, J.; Pi, S.; Zhang, Y.; Lin, T. Weak Coupling Technology with Noncoplanar Bucking Coil in a Small-Loop Transient Electromagnetic System. *IEEE Trans. Ind. Electron.* **2022**, *69*, 3151–3160. [CrossRef]
2. Lin, J.; Chen, J.; Zhang, Y. Rapid and High-Resolution Detection of Urban Underground Space Using Transient Electromagnetic Method. *IEEE Trans. Ind. Inform.* **2022**, *18*, 2622–2631. [CrossRef]
3. Spies, B.R. Depth of Investigation in Electromagnetic Sounding Methods. *Geophysics* **1989**, *54*, 872–888. [CrossRef]
4. Larsen, J.J.; Pedersen, S.S.; Foged, N.; Auken, E. Suppression of Very Low Frequency Radio Noise in Transient Electromagnetic Data with Semi-Tapered Gates. *Geosci. Instrum. Methods Data Syst.* **2021**, *10*, 81–90. [CrossRef]
5. Sun, Y.; Huang, S.; Zhang, Y.; Lin, J. An Efficient Preprocessing Method to Suppress Power Harmonic Noise for Urban Towed Transient Electromagnetic Measurement Data. *Measurement* **2021**, *186*, 110171. [CrossRef]

6. Huang, S.; Wang, S.; Sun, Y.; Zhang, Y. Efficient Processing Power Harmonic Noise with Fluctuation Frequency in Urban Transient Electromagnetic Surveys. *Rev. Sci. Instrum.* **2021**, *92*, 044501. [CrossRef] [PubMed]
7. Lin, T.; Zhou, K.; Cao, Y.; Wan, L. A Review of Air-Core Coil Sensors in Surface Geophysical Exploration. *Measurement* **2022**, *188*, 110554. [CrossRef]
8. Chen, C.; Liu, F.; Lin, J.; Zhu, K.; Wang, Y. An Optimized Air-Core Coil Sensor with a Magnetic Flux Compensation Structure Suitable to the Helicopter TEM System. *Sensors* **2016**, *16*, 508. [CrossRef] [PubMed]
9. Pi, S.; Yan, F.; Zhang, Y. Optimization and Design of Wide-Band and Low-Noise Air-Core Coil Sensor for TEM System. *IOP Conf. Ser. Earth Environ. Sci.* **2021**, *660*, 012006. [CrossRef]
10. Lin, F.; Wang, X.; Chen, K.; Hu, D.; Gao, S.; Zou, X.; Zeng, C. The Development and Test Research of a Multichannel Synchronous Transient Electromagnetic Receiver. *Geosci. Instrum. Methods Data Syst.* **2018**, *7*, 209–221. [CrossRef]
11. Wang, L.; Zhang, S.; Chen, S.; Luo, C. Fast Localization and Characterization of Underground Targets with a Towed Transient Electromagnetic Array System. *Sensors* **2022**, *22*, 1648. [CrossRef] [PubMed]
12. Gao, J. Noise Sources and Noise Characteristics of Amplifiers. In *Detection of Weak Signals*; Gao, J., Ed.; New Series of Textbooks "Information, Control Desire System"; Tsinghua University Press: Beijing, China, 2011; p. 82. ISBN 978-7-302-53067-1.
13. Wang, Y.; Shi, J.; Shi, H. Low noise optimization design of conditioning circuit for ZTEM airborne magnetic sensor. *Chin. J. Sci. Instrum.* **2018**, *39*, 187–194. [CrossRef]
14. Levinzon, F.A. Measurement of Low-Frequency Noise of Modern Low-Noise Junction Field Effect Transistors. *IEEE Trans. Instrum. Meas.* **2005**, *54*, 2427–2432. [CrossRef]
15. Scandurra, G.; Cannatà, G.; Ciofi, C. Differential Ultra Low Noise Amplifier for Low Frequency Noise Measurements. *AIP Adv.* **2011**, *1*, 022144. [CrossRef]
16. Cannatà, G.; Scandurra, G.; Ciofi, C. An Ultralow Noise Preamplifier for Low Frequency Noise Measurements. *Rev. Sci. Instrum.* **2009**, *80*, 114702. [CrossRef] [PubMed]
17. Levinzon, F.A. Ultra-Low-Noise High-Input Impedance Amplifier for Low-Frequency Measurement Applications. *IEEE Trans. Circuits Syst. Regul. Pap.* **2008**, *55*, 1815–1822. [CrossRef]
18. Lin, J.; Jia, W.; Pi, S.; Zhang, Y. Research on non-coplanar eccentric self-compensation zero-coupling receiving-transmitting technology for small size TEM. *Chin. J. Sci. Instrum.* **2020**, *8*, 150–159. [CrossRef]
19. Yu, C.; Fu, Z.; Zhang, H.; Tai, H.-M.; Zhu, X. Transient Process and Optimal Design of Receiver Coil for Small-Loop Transient Electromagnetics: Receiver Coil for Small-Loop Transient Electromagnetics. *Geophys. Prospect.* **2014**, *62*, 377–384. [CrossRef]
20. Chen, J.; Yan, F.; Sun, Y.; Zhang, Y. Applicability of transient electromagnetic fast forward modeling algorithm with small loop. *Prog. Electromagn. Res. M* **2020**, *98*, 159–169. [CrossRef]
21. Xu, Z.; Fu, Z.; Fu, N. Firefly Algorithm for Transient Electromagnetic Inversion. *IEEE Trans. Geosci. Remote Sens.* **2022**, *60*, 1–12. [CrossRef]

Article

Demonstration of 144-Gbps Photonics-Assisted THz Wireless Transmission at 500 GHz Enabled by Joint DBN Equalizer

Xiang Liu [1,2], Jiao Zhang [1,2,*], Shuang Gao [1], Weidong Tong [1,2], Yunwu Wang [1,2], Mingzheng Lei [2], Bingchang Hua [2], Yuancheng Cai [1,2], Yucong Zou [2] and Min Zhu [1,2,*]

[1] National Mobile Communications Research Laboratory, Southeast University, Nanjing 210096, China
[2] Purple Mountain Laboratories, Nanjing 211111, China
* Correspondence: jiaozhang@seu.edu.cn (J.Z.); minzhu@seu.edu.cn (M.Z.)

Abstract: The THz wireless transmission system based on photonics has been a promising candidate for further 6G communication, which can provide hundreds of Gbps or even Tbps data capacity. In this paper, 144-Gbps dual polarization quadrature-phase-shift-keying (DP-QPSK) signal generation and transmission over a 20-km SSMF and 3-m wireless 2 × 2 multiple-input multiple-output (MIMO) link at 500 GHz have been demonstrated. To further compensate for the linear and nonlinear distortions during the fiber–wireless transmission, a novel joint Deep Belief Network (J-DBN) equalizer is proposed. Our proposed J-DBN-based schemes are mainly optimized based upon the constant modulus algorithm (CMA) and direct-detection least mean square (DD-LMS) equalization. The results indicate that the J-DBN equalizer has better bit error rate (BER) performance in receiver sensitivity. In addition, the computational complexity of the J-DBN-based equalizer can be approximately 46% lower than that of conventional equalizers with similar performance. To our knowledge, this is the first time that a novel joint DBN equalizer has been proposed based on classical algorithms. It is a promising scheme to meet the demands of future fiber–wireless integration communication for low power consumption, low cost, and high capacity.

Keywords: Terahertz-band; multiple-input multiple-output; polarization multiplexing; seamless integration; Deep Belief Network

1. Introduction

The explosive growth of wireless devices and streaming services, e.g., 3D video, cloud office, virtual reality (VR), internet of vehicles, and 6G services, has led to the demand for higher transmission capacity and more spectrum resources [1,2]. The Terahertz band (THz band), occupying the spectrum range of 0.3 THz to 10 THz, is a promising candidate for providing large capacity; it can provide a data capacity of hundreds of Gbps or even Tbps due to its substantial available bandwidth [3–5]. Moreover, the devices at THz band have a small and compact size, and can be monolithically integrated with other front-end circuits in portable terminals.

THz communication techniques can be divided into two main categories: pure electronic schemes and photonics-aided schemes. Photonics-aided THz-wave signals can be adopted in fiber–wireless communication systems, which overcomes the bandwidth limitation and electromagnetic interference resulting from electronic devices [6–8]. Moreover, the characteristics of photonic technology have effectively promoted the seamless integration of wireless and optical fiber networks. A photonics-assisted transmission system at the THz band integrates the large-capacity, long-distance advantages of fiber-optic transmission and a THz wireless transmission link, and therefore it has excellent potential for future 6G communication. In previous years, several seamless integration systems of THz wireless and optical fiber networks have been demonstrated, enabled by photonics [9–15]. However, the nonlinear impairments induced by photoelectric devices and wireless links will cause severe degradation and significantly restrict the transmission rate and distance. Therefore, advanced equalization algorithms are required to compensate for these nonlinear impairments.

To mitigate the distortions of nonlinearity, several nonlinear equalization methods have been widely researched, i.e., Volterra [16], Kernel [17], and Maximum Likelihood Sequence Estimator (MLSE) [18]. However, with the increase in transmission speed, conventional digital signal processing (DSP) algorithms cannot effectively equalize nonlinear noise such as relative intensity and modal partition noise in the practical system [19]. In addition, blind equalization, which performs adaptive channel equalization without the aid of a training sequence, has become a favorite of practitioners, such as the constant modulus algorithm (CMA) equalizer and direct-detection least mean square (DD-LMS) equalizer. As we know, CMA is a typical blind adaptive algorithm for adaptive blind equalizers in Quadrature Amplitude Modulation (QAM) format [20]. Nevertheless, CMA has a significant residual error after convergence, which is unsuitable for the nonlinear channel balance. Furthermore, more advanced artificial intelligence (AI) algorithms are considered as promising solutions for signal equalization in optical communication systems. Table 1 summarizes the typical photonics-aided wireless communication systems using AI equalization algorithms. Different paradigms based on artificial neural networks (CNN) [21,22] and deep neural networks (DNN) [23–26] have been proven to obtain better performance than conventional DSP algorithms, showing an excellent capability of mitigating nonlinear distortions. However, a CNN fails to fully account for the relevant characteristic information of the input sample, leading to performance degradation in the optical communication system [27]. Moreover, a CNN is unable to meet the demand for complex-valued equalization in wireless transmission systems [24]. Furthermore, a DNN requires many training sequences in order to achieve the optimum result, reducing the effective channel capacity. Due to the ability to learn the relevant information and make the nonlinear decision in feature space, the Deep Belief Network (DBN) is used for nonlinear equalization, which is more helpful in handling the receiver sensitivity issue in the THz-band wireless communication system [28].

Table 1. Summary of photonics-aided wireless transmission systems based on AI algorithms.

Frequency (GHz)	Line Rate (Gb/s)	Format	Distance (m)	Pol. Mux	AI Algorithm	Year	Ref.
340	53.5	16QAM	54.6	Single	CNN	2022	[21]
120	40/55	16/64QAM	200	Single	2D-CNN	2022	[22]
135	60	PAM-8	3	Single	DNN	2020	[23]
140	90	PAM-4	3	Single	DNN/LSTM	2021	[24]
80	60	64QAM	1.2	Single	Dual-GRU	2022	[25]
500	144	QPSK	3	Dual	DBN	2022	This work

In this study, we propose a novel joint DBN (J-DBN) equalizer for the combination of two loss functions based on blind CMA and DD-LMS equalizers. The J-DBN nonlinear equalizer not only retains an excellent capability of mitigating nonlinear impairment but also reduces the computational complexity. By using our proposed J-DNN equalizer, we experimentally off-line demonstrated a 144 Gbit/s dual polarization QPSK (DP-QPSK) signal generation over a 20-km standard single-mode fiber (SSMF) and 3-m wireless 2 × 2 multiple-input multiple-output (MIMO) link at 500 GHz with BER below 3.8×10^{-3}. The results show that a J-DBN equalizer can significantly improve the receiver sensitivity performance. In addition, the computational complexity of the J-DBN equalizer can be approximately 46% lower than that of traditional equalizers with similar BER performance.

2. Operation Principle for J-DBN-Based Equalization

2.1. Traditional DBN Method

The DBN can be approximated as a stack of restricted Boltzmann machines (RBMs) [28], as shown in Figure 1a. An RBM is a generative stochastic network containing a visible layer $v = \{v_i\}_{i=1}^{N}$ and a hidden layer $h = \{h_j\}_{j=1}^{M}$ with the parameters $\theta = \{W, d, c\}$. The energy function and the likelihood function of the RBM can be stated as

$$E(v,h) = -\sum_{i,j} v_i W_{ij} h_j - \sum_i d_i v_i - \sum_j c_j h_j, \tag{1}$$

$$P(v,h) = \frac{1}{Z} \exp(-E(v,h)), \tag{2}$$

where $v_i \in \{0,1\}$, $h_j \in \{0,1\}$, $W = (W_{ij}) \in R^{N \times M}$ are the weights connecting the visible layer and hidden layer, d and c are the bias terms of the visible and hidden layers, and Z represents the partition function. Moreover, the probabilities $P(v|h)$ and $P(h|v)$ can be calculated by

$$P(v_i = 1|h) = \sigma\left(\sum_j W_{ij} h_j + d_i\right), \tag{3}$$

$$P(h_j = 1|v) = \sigma\left(\sum_i W_{ij} v_i + c_j\right), \tag{4}$$

where $\sigma(\cdot)$ is the sigmoid function defined as $\sigma(x) = 1/(1 + e^{-x})$.

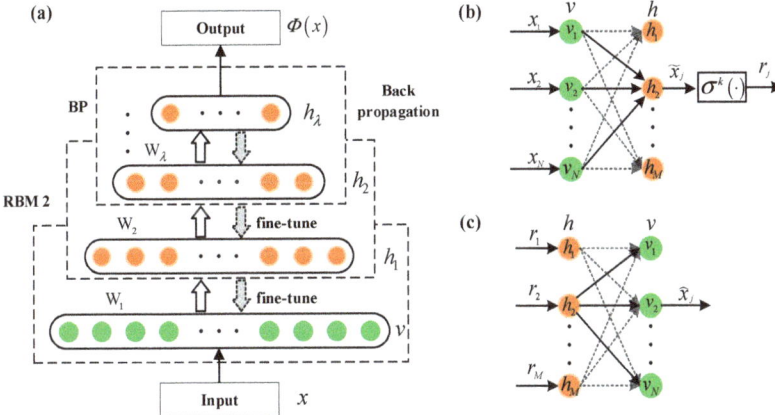

Figure 1. (a) The structure of DBN, which can be approximated as a stack of RBMs. (b) The feedforward architecture of the j-th hidden neuron. (c) The reconstruction architecture of the j-th hidden neuron.

The DBN with λ hidden layers contains $W_1, W_2, \ldots, W_\lambda$ connection weight matrices and $\lambda + 1$ biases $d_0, d_1, \ldots, d_\lambda$, where d_0 is the bias of the visible layer v. Therefore, the output probability can be calculated by the hidden vector: $\Phi = \sigma(W_\lambda h_{\lambda-1} + d_\lambda)$.

Figure 1b,c represent the j-th neuron of the feedforward and reconstruction architecture. The RBM utilizes the stochastic approximation method to update the parameters $\theta = \{W, d, c\}$ by maximizing the likelihood $P_\theta(v)$.

2.2. J-DBN Nonlinear Equalizers

There are two main NN-based equalization schemes, namely the blind NN equalization and adaptive NN equalization algorithms. To avoid the residual error for equalizing nonlinear channels, the CMA equalizer can be combined with the NN algorithm. Meanwhile, DD-LMS combined with the NN equalization algorithm can further accelerate the convergence speed and ensure the optimum nonlinear decision. Our proposed J-DBN method, including an adaptive DBN-1-based equalizer and a blind DBN-2-based equalizer, is mainly based upon CMA and DD-LMS blind equalizers, as shown in Figure 2a.

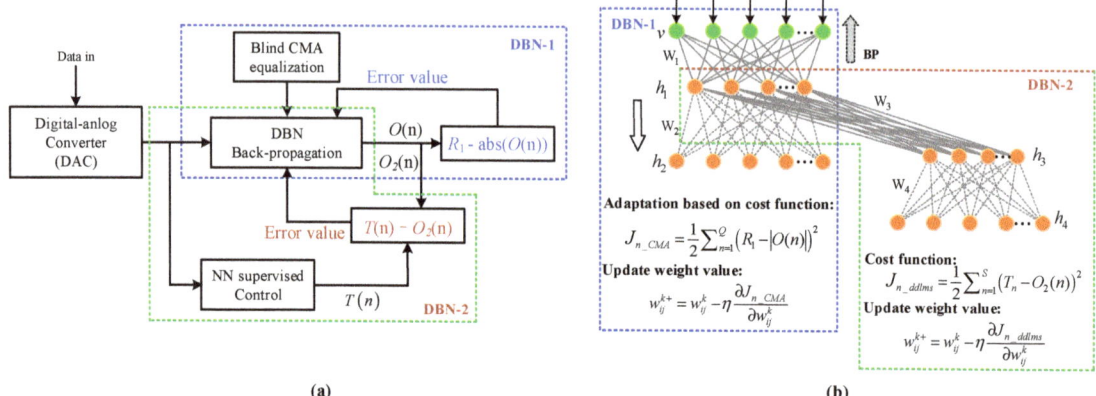

Figure 2. The proposed J-DBN equalizers: (**a**) schematic diagram; (**b**) architecture.

Figure 2b shows the detailed architecture of our proposed J-DBN equalizer, including two sequential DBNs. Each link between one visible layer and multiple hidden layers in J-DBN is associated with the weight value w_{ij}^k, where k denotes the k-th hidden layer, and i and j represent the i-th node in the visible layer and the j-th node in the current hidden layer, respectively. The output of nonlinear neurons is summed as $h_j^k = \sigma\left(\sum_{i=1}^S w_{ij}^k x_i\right)$, where S is defined as the length of input samples, and $\sigma(\cdot)$ denotes the nonlinear active function between multiple hidden layers. The selection of a matching activation function is an important part of DBN construction.

During the feedforward training process, the output of the j-th neuron in the other $(k+1)$-th visible layer can be calculated as $v_j^{k+1} = \sigma\left(\sum_{i=1}^N w_{ij}^{k+1} v_j^k\right)$, where v_j^k is the output of the j-th neuron cell in the kth hidden layer, and the total number of cells in this layer is N. When k is increased as λ, we define h_j^{L+1} as the final output signal calculated from the output layer. The DBN-1 and DBN-2 equalizers update the weights according to their respective cost functions, guaranteeing the optimal solution of the whole equalization process.

2.3. Cost Function of DBN-1 and DBN-2 Equalizers
2.3.1. Cost Function of the Adaptive DBN-1-Based Equalizer

As we know, CMA is a typical blind equalizer for OOK or QPSK modulation format with a single reference radius. To reduce the large residual error after convergence, CMA blind equalization can be combined with a DBN-1 equalizer. The corresponding filter tap weight h_{nn} and the cost function of the DBN-1 equalizer based on the CMA algorithm are defined as

$$h_{nn} \to h_{nn} + \mu e_{n\_CMA} x(n) \quad (5)$$

$$J_{n\_CMA} = \frac{1}{2}\sum_{n=1}^Q (R_1 - |O(n)|)^2 \quad (6)$$

where μ is defined as the CMA convergence parameter, R_1 is the constant module of signals, $O(n)$ is the output of the DBN-1 network, and the input sequence $x(n)$ in the optimization network is the test signal with a size of Q.

The adaptive error function can be defined as $e_{n\_CMA} = R_1 - |O(n)|$. In practical terms, we can find the optimum μ and tap number to obtain better channel equalization. Moreover, the weight value can be further optimized with the aid of adaptive equalization and the BP algorithm, which can be given as

$$w_{ij}^{k+} = w_{ij}^k - \Delta w_{ij}^k = w_{ij}^k - \eta \frac{\partial J_{n\_CMA}}{\partial w_{ij}^k} \quad (7)$$

2.3.2. Cost Function of Blind DBN-2-Based Equalizer

Unlike forwarding propagation steps, the weight values and model hyper-parameters are updated based on the backpropagation (BP) algorithm. According to the minimum mean square error (MMSE) algorithm, the cost function of the DBN-2 blind equalizer based on the DD-LMS algorithm can be defined as

$$J_{n\_ddlms} = \frac{1}{2} \sum_{n=1}^{S} (T_n - O_2(n))^2 \quad (8)$$

Compared the obtained value $O_2(n)$ with the corresponding expected value T_n, the error $e_n = T_n - O_2(n)$ is sent to the network with a reverse training algorithm. Thus, the connected weight vector w_{ij}^{k+} can be iteratively updated until the desired epoch or error value is reached, which can be represented as

$$w_{ij}^{k+} = w_{ij}^k - \Delta w_{ij}^k = w_{ij}^k - \eta \frac{\partial J_{n\_ddlms}}{\partial w_{ij}^k} \quad (9)$$

where η is the learning rate and Δ represents the gradient operation. In order to accurately update the weight value of every nonlinear node, we must calculate the gradient of the whole training sequence. Owing to the introduction of the blind error function, we will decrease the training size and improve the computation speed effectively.

Thus, our proposed J-DBN equalizer comprises two parts, including the DBN-1 adaptive equalization and the DBN-2 blind equalization. Note that the former updating of the weight value is based on the traditional BP algorithm, while the latter deploys the blind equalization algorithm to optimize the network further.

3. Experimental Setup

In our previous work, we have established some real-time photonics-aided THz seamless integration transmission systems [12–15]. In order to verify the effectiveness of our proposed algorithms in this paper, we further perform an experimental demonstration for 144−Gbps photonics-assisted THz wireless transmission at 500 GHz enabled by J-DBN equalization. A detailed description of the experimental setup is shown in Figure 3, including the optical and THz transmitter modules, THz 2 × 2 MIMO wireless link, THz receiver module, and the off-line DSP blocks. For a fair comparison, there were four alternative algorithms included in the experiment at the off-line DSP blocks.

3.1. Optical and THz Transmitter Modules

Figure 3 provides the experimental setup of the optical and THz transmitter modules. An arbitrary waveform generator (AWG) of 92-Gsa/s sampling rate is used to generate the I and Q components of the baseband electrical signals. A parallel electrical amplifier (EAs) is used to amplify the I/Q electrical signals. Then, a 193.5-THz linewidth external cavity laser-1 (ECL-1) is used to produce the optical carrier, which is modulated via an I/Q modulator with 30-GHz bandwidth and approximately 7-dB insertion loss. The modulated optical signal is divided to the polarization multiplexing channels by an optical coupler (OC) to simulate signal delay and attenuation. One is transmitted through a 1-m fiber direct link (DL) and the other passes through a variable optical attenuator (VOA). Then, after a polarization beam coupler (PBC), the 193.5-THz optical baseband signal is sent to SSMF.

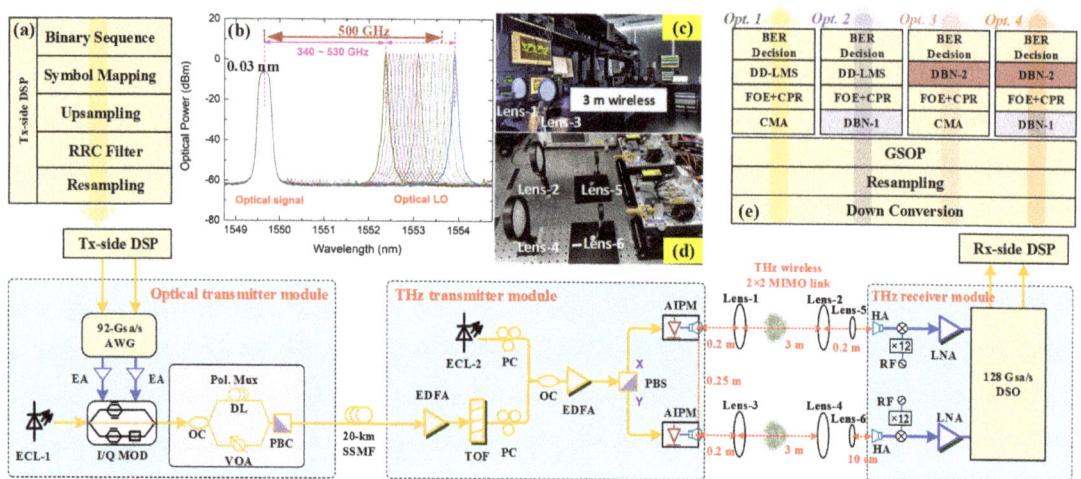

Figure 3. Experimental setup of photonics–aided THz wireless transmission system over 20-km SSMF and 3-m wireless distance with detailed DSP blocks at Tx- and Rx-side, including (**a**) Tx-side DSP block, (**b**) the optical spectra of the optical signal with tunable optical LO after optical coupler (0.03 nm resolution). Insets: (**c**) 3-m 2 × 2 MIMO wireless transmission link. (**d**) Lens position at THz receiver side. (**e**) Rx-side DSP block with three proposed J–DBN optional schemes.

Erbium-doped fiber amplifiers (EDFA) are used to compensate for the optical-fiber transmission loss after a 20-km SSMF transmission link. In order to suppress the out-of-band amplified spontaneous emission (ASE) noise, a passband tunable optical filter (TOF) is used. A free-running tunable external cavity laser (ECL-2) is operated as an optical local oscillator (LO), which has a linewidth of less than 100 kHz. An optical baseband signal with 10.5 dBm optical power and an optical LO with 13.5 dBm optical power are coupled by an OC. The optical spectra of the optical signal with the tunable optical LO after OC (0.03 nm resolution) are shown in Figure 3b. Note that the optical power of the X- and Y-polarization components after PBS should be as equal as possible. The AIPMs used in our setup are polarization-sensitive, with a maximum of 4.5 dB polarization-dependent responsivity (PDR). The X- and Y-polarization imbalance will result in 2 × 2 MIMO THz-wave imbalance and deteriorate the system performance. Therefore, two polarization controllers (PCs) are required before the OC. In the test, the optical signal and optical LO separately adjust the incident X- and Y-polarization direction to maximize the optical power in the antenna-integrated photomixer module (AIPM, NTT Electronics Corp. IOD-PMAN-13001).

The side-mode suppression ratio (SMSR) of the optical signal and optical LO is >50 dB. In our proposed system, THz-wave wireless signals with a tunable carrier frequency range from 340 GHz to 530 GHz are generated by photonic heterodyning using AIPMs. The AIPM consists of an ultra-fast uni-traveling-carrier photodiode (UTC-PD) and a bow-tie or log-periodic antenna. Two parallel AIPMs are used, each with a typical −28 dBm output power and operating wavelength range from 1540 nm to 1560 nm. The typical photodiode responsivity is 0.15 A/W, and the maximum optical input power is 15 dBm. PBS is used to separate the X- and Y-polarization components by APIMs to generate two parallel THz-wave wireless signals, respectively. To drive the AIPMs, another EDFA is used to boost the optical power of combined lightwaves before PBS.

3.2. THz 2 × 2 MIMO Wireless Link and THz Receiver Modules

Figure 3 shows the THz 2 × 2 MIMO wireless transmission link and lens position. The two parallel THz-wave signals from the AIPMs are transmitted over a 3-m 2 × 2 MIMO wireless transmission link. In order to focus the wireless THz-wave, three pairs of lenses are deployed to maximize the received THz-wave signal power and are manually aligned. Lenses 1, 2, 5 and lenses 3, 4, 6 are aligned with the X-polarization and Y-polarization wireless link, respectively. Lenses 1–4 are identical, each having a 20-cm focal length and 10-cm diameter. The smaller lenses 5 and 6 have a 10-cm focal length and 5-cm diameter, which are used for THz-waves' high-accuracy alignment to horn antennas (HA). For the X-polarization (Y-polarization) THz wireless link, the longitudinal separation distance between the AIPM and lens 1 (lens 3), lens 2 (lens 4), and lens 5 (lens 6) and the receiver HA are 0.2 m, 3 m, and 5 cm, respectively. The lateral separation between the two AIPMs and two HA pairs is 25 cm. In order to avoid multi-path fading from reflections on the optical table, the conversion modules are placed at the height of 20 cm. Photos of the 3-m wireless transmission link and lens position are shown in Insets Figure 3c and Figure 3d, respectively.

At the THz-wave receiver end, THz-wave wireless signals are received with two parallel THz-band HAs with 26-dBi gain. Electronic LO sources drive two identical THz receivers to implement analog down-conversion for X- and Y-polarization THz-wave wireless signals. Each consists of a mixer, a ×12 frequency multiplier chain, and an amplifier, and operates at 500 GHz. Then, the down-converted X- and Y-polarization intermediate-frequency (IF) signal at 20 GHz is boosted by two cascaded electrical low-noise amplifiers (LNAs) with a 3-dB bandwidth of 47 GHz and captured by a digital oscilloscope with a 128-Gsa/s sampling rate.

3.3. Off-Line DSP Blocks

The block diagram of the off-line DSP is illustrated in Figure 3a,e. In the Tx-side DSP, a QPSK symbol mapping and a raised-cosine (RC) filter with a roll-off factor of 0.01 are deployed, as shown in Figure 3a. Figure 3e depicts the four DSP options at the Rx side, verifying the validity of the J-DBN-based equalization schemes. Conventional, Opt. 1: the captured signal is offline, proposed by typical DSP steps including down-conversion into baseband, resampling, Gram–Schmidt orthogonalization process (GSOP), followed by 53-tap CMA equalization. Then, the frequency offset noise can be mitigated via frequency offset estimation (FOE), and the phase offset problem can be solved after carrier phase recovery (CPR). Finally, a 37-tap DD-LMS equalizer is added to compensate for the remaining linear damage and I/Q imbalance before BER calculation.

We also compare the BER performance between the CMA equalizer, DBN-based equalizer, and J-DBN equalizer within these DSP steps. In Opt. 2, the adaptive DBN-1 equalizer adapts itself to compensate for the nonlinear distortion. It extracts the signal sequences' characteristics based on the BP algorithm and blind CMA algorithm. The scheme can reduce the sizeable residual error after convergence, which is suitable for the nonlinear channel balance. In Opt. 3, the blind DBN-2 equalizer optimizes the weight value and tap number based on the BP algorithm and DD-LMS equalizer. Moreover, the DBN-2 equalizer can utilize the weight value updated by the DBN-1 as the initial value. The scheme can further compensate for the remaining linear damage and nonlinear decision ability. In Opt. 4, the J-DBN equalizer combines two error cost functions, which is more helpful in handling the receiver sensitivity issue in THz-band wireless links and achieving a more accurate BER decision. Moreover, the J-DBN equalizer can be established via two steps. Firstly, it can be initialized with the aid of the training sequence, and then the weight value can be further optimized by employing the error function of CMA. Adopting our proposed J-DBN equalizer, both the lengths of the training data and the training time can be effectively reduced.

4. Experimental Results and Discussion

As we know, a well-designed equalizer is useful for resolving nonlinear issues and has been successfully applied in wireless communications. However, the selection of an error function is an important factor that affects the residual error of the blind equalizer; thus, the performance of the equalizer is different. In our experiment, we introduce some equalizers in our proposed schemes and compare their performance, such as the typical CMA equalizer with taps and the DD-LMS equalizer.

4.1. Transmission Results

We first measure the performance of our transparent fiber-optical and THz wireless 2×2 MIMO transmission system for the back-to-back (BtB) case, i.e., without fiber and wireless distance transmission. Figure 4 gives the measured BER of X- and Y-polarization versus different input power into each AIPM. At the 3.8×10^{-3} HD-FEC limit, the THz-wave carrier frequency at 500 GHz for the BtB case has a successfully transmitted range from 28 GBaud to 36 GBaud. The insets show that the QPSK constellation points can be demodulated well at 28 GBaud. However, the BER performance degrades at high transmission rates due to the limitations of the receiver LNA bandwidth.

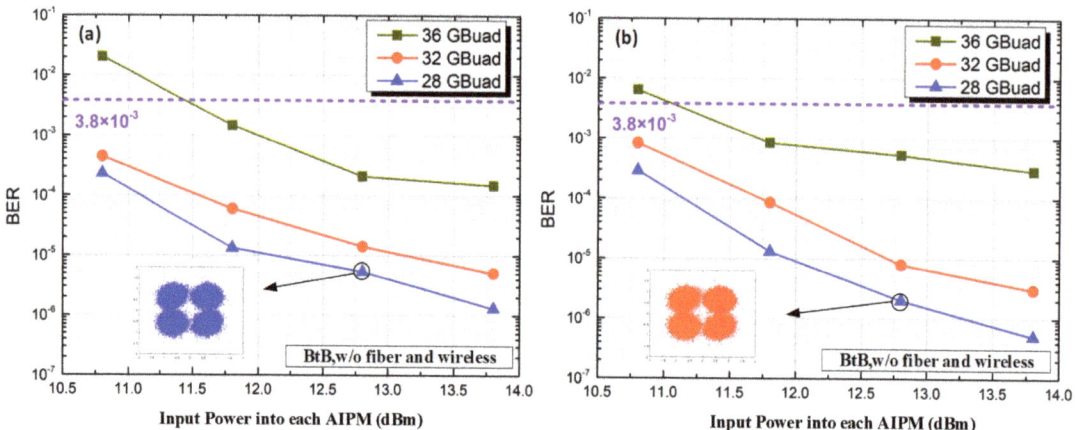

Figure 4. BER versus input power into each AIPM for the BtB case without fiber and wireless transmission. (**a**) X-polarization; (**b**) Y-polarization.

Then, we measure the BER versus the input power into each AIPM over one span of 20-km SSMF and 3-m wireless distance, as shown in Figure 5. At the 3.8×10^{-3} HD-FEC limit, the THz-wave carrier frequency at 500 GHz for the fiber and wireless transmission can also be successfully transmitted. However, the transmission performance is not ideal at 36 GBaud. Figure 6 gives the electrical spectrum of the 24/28/32/36 GBaud QPSK IF signal with the corresponding bandwidth (BW). The signals are more damaged when the transmission speed increases. Moreover, we can see that the required bandwidth becomes limited as the transmission rate increases, especially at the 36 GBaud rate with 36.36 GHz bandwidth. To protect the AIPMs, the maximum input optical power is set at 13.5 dBm. A THz-wave carrier frequency at 500 GHz can be successfully transmitted at the 3.8×10^{-3} HD-FEC threshold. The best BER performance occurs for the lower transmission rates. The insets also show that the constellation points can be demodulated well. In order to further improve the performance for higher transmission rates, the J-DBN equalizer based on the conventional DSP algorithm is introduced.

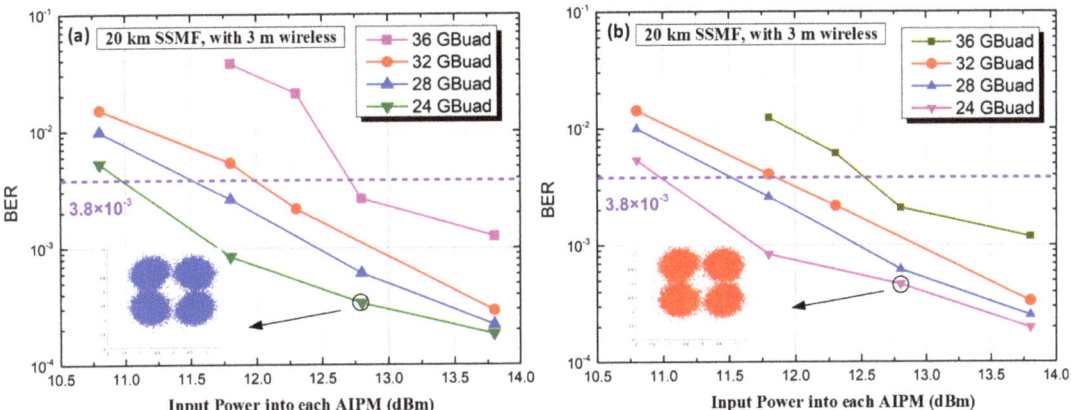

Figure 5. BER versus input power into each AIPM over one span of 20-km SSMF and 3-m wireless transmission. (**a**) X-polarization; (**b**) Y-polarization.

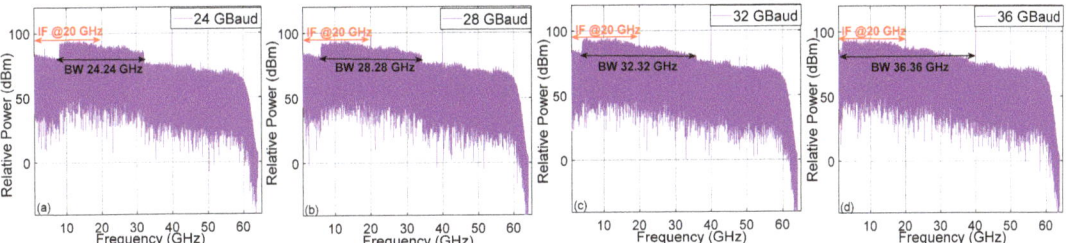

Figure 6. Electrical spectrum of the received QPSK IF signal: (**a**) 24 GBaud signal with 24.24 GHz BW; (**b**) 28 GBaud signal with 28.28 GHz BW; (**c**) 32 GBaud signal with 32.32 GHz BW; (**d**) 36 GBaud signal with 36.36 GHz BW.

Here, the training data are only used for the DBN-1 adaptive equalizer in our proposed J-DBN equalizer since the DBN-2 blind equalizer and optimization is a self-recovering equalization method without the aid of a training sequence. It indicates that the J-DBN equalizer has good training accuracy and satisfactory tracking speed. Figure 7 illustrates the BER of 36 GBaud QPSK signals versus the input optical power into each AIPM and SNR over one span of 20-km SSMF and 3-m wireless distance; it can be found that increasing the optical power can help to improve the BER performance due to the larger SNR. Next, we compare four equalization schemes (including *Opt. 1–Opt. 4*) with a 53-tap CMA equalizer, 37-tap DD-LMS equalizer, DBN-1 equalizer with 25 cells, and DBN-2 equalizer with 50 cells. From the comparison between *Opt.* 1 and *Opt.* 4, it can be concluded that the required power under HD-FEC (3.8×10^{-3}) utilizing the J-DBN scheme is close to 12.6 dBm, which increases by almost up to 0.2 dB in receiver sensitivity and 0.8 dB in SNR gain compared with the conventional scheme. Moreover, *Opt.* 2 and *Opt.* 3 also obtain a slight performance improvement after the DBN-1 equalizer or DBN-2 equalizer. When the input power into the AIPM is 12.8 dBm, the illustrations depict the constellation diagrams of QPSK symbols after recovery. The received QPSK symbol constellation before the receiver DSP chain is also given. Furthermore, we compared the constellation diagrams of the *Opt.* 1 and *Opt.* 4 schemes. The illustrations (i) and (ii) also show that the J-DBN equalizer can reduce the residual error after convergence and visually improve the nonlinear decision capacity, which is suitable for the nonlinear channel balance.

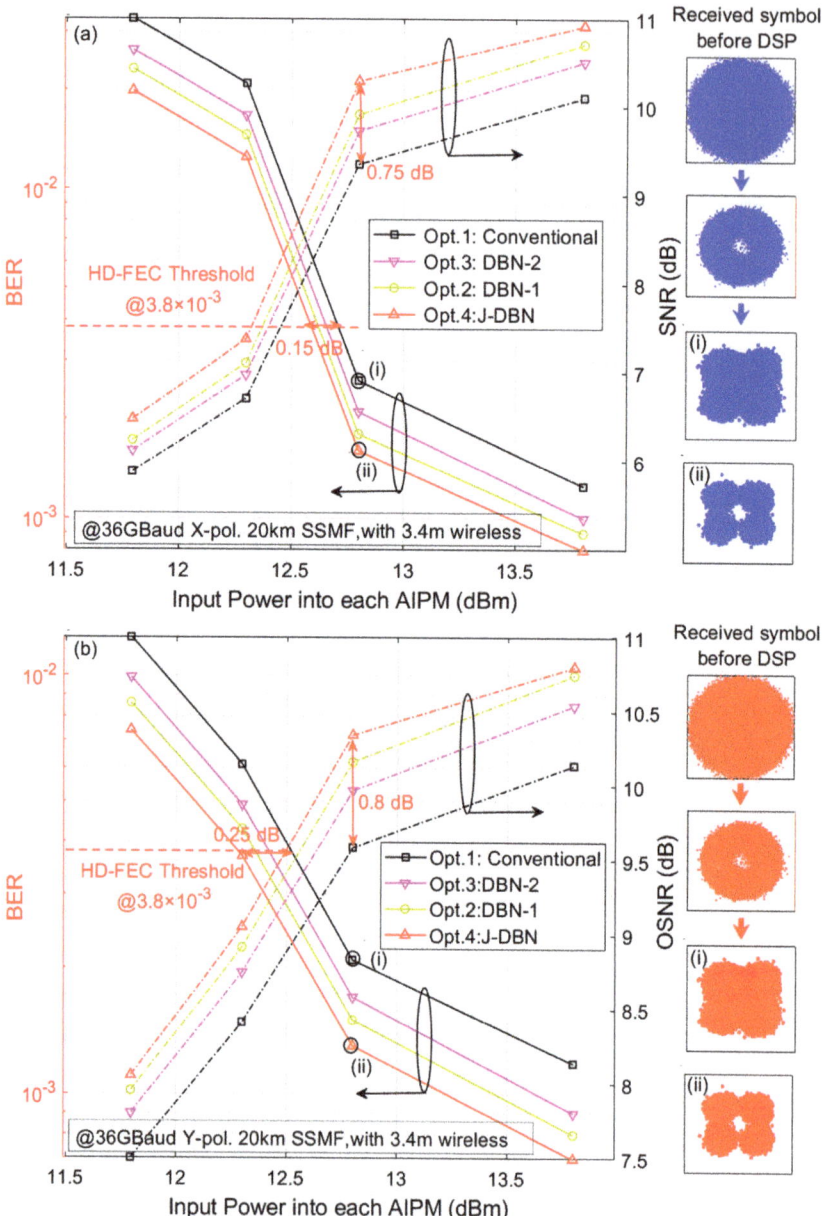

Figure 7. BER versus input power into each AIPM over one span of 20-km SSMF and 3-m wireless transmission for different proposed equalizers. (**a**) X-polarization. (**b**) Y-polarization.

4.2. Complexity Analysis

We further analyze the complexity of the proposed J-DBN nonlinear equalizer and make a comparison with blind CMA and DD-LMS equalizers. We consider the complexity in two aspects, convergence steps and computation time, as is shown in Figure 8. Figure 8a shows that the convergence speeds of the DBN-1 and DBN-2 equalizers are faster than the conventional CMA and DD-LMS equalizers, which verifies the well-trained neural networks. Moreover, the computation time of DBN-based equalizers can be approximately 36.3% and 46% lower than that of CMA and DD-LMS-based methods with similar BER performance, as is shown in Figure 8b. Therefore, the conclusion can be reached that the J-DBN equalizer has a significant advantage in dealing with nonlinear distortion, making it quite suitable in high-speed THz-band wireless communication systems.

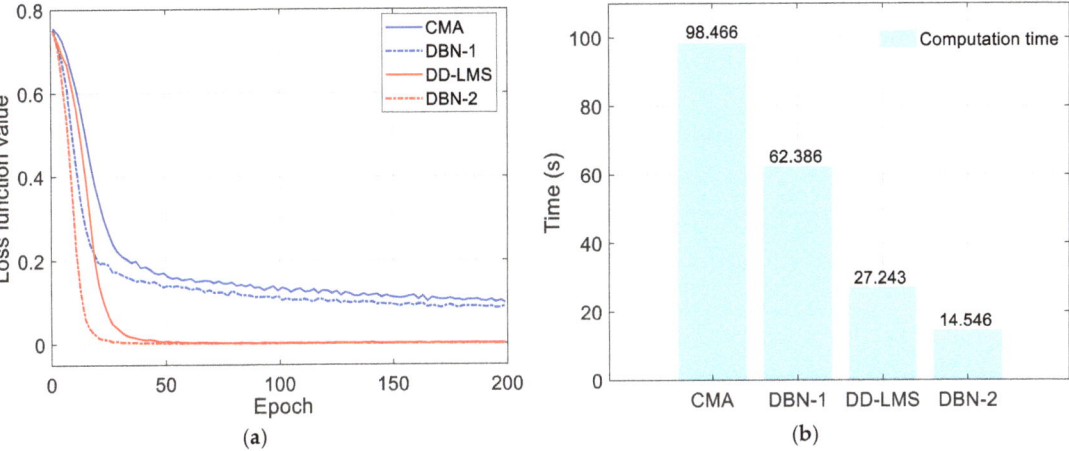

Figure 8. Complexity analysis of the 36 Gbaud QPSK signals with different equalization options when the input power into AIPM is 12.8 dBm. (**a**) Loss function versus iteration numbers. (**b**) Comparison of the computation time.

5. Conclusions

In this paper, a novel J-DBN equalizer for the 144-Gbps QPSK signal transmission system over a 20-km SSMF and 3-m THz-band wireless link at 500-GHz is experimentally demonstrated, which consists of two steps including a DBN-1 adaptive equalizer based on the CMA algorithm and a DBN-2 blind equalizer based on the DD-LMS algorithm. We compare the J-DBN equalizer with the classical equalizer in terms of the BER performance. Meanwhile, our proposed J-DBN equalizer in the adaptive equalization step can reduce the computational complexity and obtain better training accuracy during the self-recovering blind equalization process. The experimental results show that the J-DBN method has good training accuracy, a smaller requirement for training sequences, and satisfactory computational complexity. Thanks to our proposed J-DBN scheme, an improvement of 0.2 dB and 0.8 dB over the J-DBN equalizer in receiver sensitivity and SNR gains at a BER of 3.8×10^{-3} is achieved compared with conventional equalizers. Moreover, the computational time is effectively improved, being up to 46% lower than the conventional equalizer. To our knowledge, this is the first time that joint DBN equalizers have been deployed in the THz-band wireless transmission link. Our proposed J-DBN equalization scheme is promising for future 6G fiber-optical and THz-wireless seamless integration communication systems.

Author Contributions: Conceptualization, X.L. and J.Z.; methodology, X.L. and J.Z.; software, W.T. and Y.W.; validation, B.H., M.L. and Y.C.; formal analysis, X.L., S.G. and J.Z.; investigation, X.L. and J.Z.; resources, X.L.; data curation, X.L.; writing—original draft preparation, X.L.; writing—review and editing, X.L., Y.Z. and J.Z.; supervision, M.Z.; project administration, M.Z. All authors have read and agreed to the published version of the manuscript.

Funding: This research was supported in part by the National Natural Science Foundation of China (62101121 and 62101126), the project funded by the China Postdoctoral Science Foundation (2021M702501 and 2022T150486), the open project (2022GZKF003) of the State Key Laboratory of Advanced Optical Communication Systems and Networks of Shanghai Jiao Tong University, the Transformation Program of Scientific and Technological Achievements of Jiangsu Province (BA2019026), the Key Research and Development Program of Jiangsu Province (BE2020012), and the Major Key Project of Peng Cheng Laboratory (PCL 2021A01-2).

Data Availability Statement: Not applicable.

Conflicts of Interest: The authors declare no conflict of interest.

References

1. You, X.; Huang, Y.; Liu, S.; Wang, D.; Ma, J.; Xu, W.; Zhang, C.; Zhan, H.; Zhang, C.; Zhang, J.; et al. Toward 6G TK u Extreme Connectivity: Architecture, Key Technologies and Experiments. Available online: https://arxiv.org/abs/2208.01190 (accessed on 2 August 2022).
2. Zhu, M.; Zhang, J.; Hua, B.C. Ultra-wideband fiber THz-fiber seamless integration communication system toward 6G: Architecture, key techniques, and testbed implementation. *Sci. China Inf. Sci.* **2022**. [CrossRef]
3. Jia, S.; Zhang, L.; Wang, S.; Li, W.; Qiao, M.; Lu, Z.; Idrees, N.M.; Pang, X.; Hu, H.; Zhang, X.; et al. 2× 300 Gbit/s line rate PS-64QAM-OFDM THz photonic-wireless transmission. *J. Lightwave Technol.* **2020**, *38*, 4715–4721. [CrossRef]
4. Zhang, H.; Zhang, L.; Wang, S.; Lu, Z.; Yang, Z.; Liu, S.; Qiao, M.; He, Y.; Pang, X.; Zhang, X.; et al. Tbit/s multi-dimensional multiplexing THz-over-fiber for 6G wireless communication. *J. Lightwave Technol.* **2021**, *39*, 5783–5790. [CrossRef]
5. Harter, T.; Füllner, C.; Kemal, J.N.; Ummethala, S.; Steinmann, J.L.; Brosi, M.; Hesler, J.L.; Bründermann, E.; Müller, A.S.; Freude, W.; et al. Generalized Kramers–Kronig receiver for coherent terahertz communications. *Nat. Photonics* **2020**, *14*, 601–606. [CrossRef]
6. Kawanishi, T. THz and photonic seamless communications. *J. Lightwave Technol.* **2019**, *37*, 1671–1679. [CrossRef]
7. Li, X.; Yu, J.; Chang, G.K. Photonics-aided millimeter-wave technologies for extreme mobile broadband communications in 5G. *J. Lightwave Technol.* **2020**, *38*, 366–378. [CrossRef]
8. Sengupta, K.; Nagatsuma, T.; Mittleman, D.M. Terahertz integrated electronic and hybrid electronic–photonic systems. *Nat. Electron.* **2018**, *1*, 622–635. [CrossRef]
9. Castro, C.; Elschner, R.; Merkle, T.; Schubert, C.; Freund, R. 100 Gb/s real-time transmission over a THz wireless fiber extender using a digital-coherent optical modem. In Proceedings of the 2020 Optical Fiber Communication Conference (OFC), San Diego, CA, USA, 8–12 March 2020.
10. Horst, Y.; Blatter, T.; Kulmer, L.; Bitachon, B.I.; Baeuerle, B.; Destraz, M.; Heni, W.; Koepfli, S.; Habegger, P.; Eppenberger, M.; et al. Transparent optical-THz-optical Link transmission over 5/115 m at 240/190 Gbit/s enabled by plasmonics. In Proceedings of the 2021 Optical Fiber Communications Conference (OFC), Francisco, CA, USA, 6–10 June 2021.
11. Horst, Y.; Blatter, T.; Kulmer, L.; Bitachon, B.I.; Baeuerle, B.; Destraz, M.; Heni, W.; Koepfli, S.; Habegger, P.; Eppenberger, M.; et al. Transparent Optical-THz-Optical Link at 240/192 Gbit/s Over 5/115 m Enabled by Plasmonics. *J. Lightwave Technol.* **2022**, *40*, 1690–1697. [CrossRef]
12. Zhang, J.; Zhu, M.; Lei, M.; Hua, B.; Cai, Y.; Zou, Y.; Tian, L.; Li, A.; Huang, Y.; Yu, J.; et al. Demonstration of Real-time 125.516 Gbit/s Transparent Fiber-THz-Fiber Link Transmission at 360~430 GHz based on Photonic Down-Conversion. In Proceedings of the 2022 Optical Fiber Communication Conference (OFC), San Diego, CA, USA, 6–10 March 2022.
13. Zhang, J.; Zhu, M.; Lei, M.; Hua, B.; Cai, Y.; Zou, Y.; Tian, L.; Li, A.; Wang, Y.; Huang, Y.; et al. Real-time demonstration of 103.125-Gbps fiber–THz–fiber 2× 2 MIMO transparent transmission at 360–430 GHz based on photonics. *Opt. Lett.* **2022**, *47*, 1214–1217. [CrossRef] [PubMed]
14. Zhang, J.; Zhu, M.; Hua, B.; Lei, M.; Cai, Y.; Zou, Y.; Tong, W.; Ding, J.; Tian, L.; Ma, L.; et al. Real-time Demonstration of 100 GbE THzwireless and Fiber Seamless Networks. *J. Lightwave Technol.* **2022**; *Early Access*.
15. Zhang, J.; Zhu, M.; Hua, B.; Lei, M.; Cai, Y.; Zou, Y.; Tian, L.; Li, A.; Huang, Y.; Yu, J.; et al. 6G Oriented 100 GbE Real-time Demonstration of Fiber-THz-Fiber Seamless Communication Enabled by Photonics. In Proceedings of the 2022 Optical Fiber Communications Conference and Exhibition (OFC), San Diego, CA, USA, 6–10 March 2022.
16. Zhang, L.; Hong, X.; Pang, X.; Ozolins, O.; Udalcovs, A.; Schatz, R.; Guo, C.; Zhang, J.; Nordwall, F.; Engenhardt, K.M.; et al. Nonlinearity-aware 200 Gbit/s DMT transmission for C-band short-reach optical interconnects with a single packaged electro-absorption modulated laser. *Opt. Lett.* **2018**, *43*, 182–185. [CrossRef] [PubMed]
17. Stojanovic, N.; Prodaniuc, C.; Zhang, L.; Wei, J. 210/225 Gbit/s PAM-6 transmission with BER below KP4-FEC/EFEC and at least 14 dB link budget. In Proceedings of the European Conference on Optical Communication (ECOC), Rome, Italy, 23–27 September 2018.
18. Zhang, L.; Pang, X.; Udalcovs, A.; Ozolins, O.; Lin, R.; Yin, X.; Tang, M.; Tong, W.; Xiao, S.; Chen, J. Kernel mapping for mitigating nonlinear impairments in optical short-reach communications. *Opt. Express* **2019**, *27*, 29567–29580. [CrossRef] [PubMed]
19. Lavrencik, J.; Pavan, S.K.; Thomas, V.A.; Ralph, S.E. Noise in VCSEL-based links: Direct measurement of VCSEL transverse mode correlations and implications for MPN and RIN. *J. Lightwave Technol.* **2016**, *35*, 698–705. [CrossRef]
20. Fijalkow, I.; Manlove, C.E.; Johnson, C.R. Adaptive fractionally spaced blind CMA equalization: Excess MSE. *IEEE Trans. Signal Process.* **1998**, *46*, 227–231. [CrossRef]
21. Wang, C.; Wang, K.; Tan, Y.; Wang, F.; Sang, B.; Li, W.; Zhou, W.; Yu, J. High-Speed Terahertz Band Radio-Over-Fiber System Using Hybrid Time-Frequency Domain Equalization. *IEEE Photonics Technol. Lett.* **2022**, *34*, 559–562. [CrossRef]

22. Wang, K.; Wang, C.; Li, W.; Wang, Y.; Ding, J.; Liu, C.; Kong, M.; Wang, F.; Zhou, W.; Zhao, F.; et al. Complex-Valued 2D-CNN Equalization for OFDM Signals in a Photonics-Aided MMW Communication System at the D-Band. *J. Lightwave Technol.* **2022**, *40*, 2791–2798. [CrossRef]
23. Zhou, W.; Zhao, L.; Zhang, J.; Wang, K.; Yu, J.; Chen, Y.W.; Shen, S.; Shiu, R.K.; Chang, G.K. 135-GHz D-band 60-Gbps PAM-8 wireless transmission employing a joint DNN equalizer with BP and CMMA. *J. Lightwave Technol.* **2020**, *38*, 3592–3601. [CrossRef]
24. Zhou, W.; Shi, J.; Zhao, L.; Wang, K.; Wang, C.; Wang, Y.; Kong, M.; Wang, F.; Liu, C.; Ding, J.; et al. Comparison of Real-and Complex-Valued NN Equalizers for Photonics-Aided 90-Gbps D-band PAM-4 Coherent Detection. *J. Lightwave Technol.* **2021**, *39*, 6858–6868. [CrossRef]
25. Liu, C.; Wang, C.; Zhou, W.; Wang, F.; Kong, M.; Yu, J. 81-GHz W-band 60-Gbps 64-QAM wireless transmission based on a dual-GRU equalizer. *Opt. Express* **2022**, *30*, 2364–2377. [CrossRef] [PubMed]
26. Chuang, C.Y.; Wei, C.C.; Lin, T.C.; Chi, K.L.; Liu, L.C.; Shi, J.W.; Chen, Y.K.; Chen, J. Employing deep neural network for high speed 4-PAM optical interconnect. In Proceedings of the 2017 European Conference on Optical Communication (ECOC), Gothenburg, Sweden, 17–21 September 2017.
27. Goodfellow, I.; Bengio, Y.; Courville, A. *Deep Learning*; MIT Press: Cambridge, MA, USA, 2016.
28. Tian, F.; Yang, C. Deep belief network-hidden Markov model based nonlinear equalizer for VCSEL based optical interconnect. *Sci. China Inf. Sci.* **2020**, *63*, 160406. [CrossRef]

Article

A W-Band Communication and Sensing Convergence System Enabled by Single OFDM Waveform

Nazar Muhammad Idrees [1], Zijie Lu [1], Muhammad Saqlain [1], Hongqi Zhang [1], Shiwei Wang [1], Lu Zhang [1] and Xianbin Yu [1,2,*]

1. College of Information Science and Electronic Engineering, Zhejiang University, Hangzhou 310027, China; nazar@zju.edu.cn (N.M.I.); 3150102348@zju.edu.cn (Z.L.); saqlain@zju.edu.cn (M.S.); zhanghongqi@zju.edu.cn (H.Z.); wsw@zju.edu.cn (S.W.); zhanglu1993@zju.edu.cn (L.Z.)
2. Zhejiang Laboratory, Hangzhou 311121, China
* Correspondence: xyu@zju.edu.cn

Abstract: Convergence of communication and sensing is highly desirable for future wireless systems. This paper presents a converged millimeter-wave system using a single orthogonal frequency division multiplexing (OFDM) waveform and proposes a novel method, based on the zero-delay shift for the received echoes, to extend the sensing range beyond the cyclic prefix interval (CPI). Both simulation and proof-of-concept experiments evaluate the performance of the proposed system at 97 GHz. The experiment uses a W-band heterodyne structure to transmit/receive an OFDM waveform featuring 3.9 GHz bandwidth with quadrature amplitude modulation (16-QAM). The proposed approach successfully achieves a range resolution of 0.042 m and a speed resolution of 0.79 m/s with an extended range, which agree well with the simulation. Meanwhile, based on the same OFDM waveform, it also achieves a bit-error-rate (BER) 10^{-2}, below the forward error-correction threshold. Our proposed system is expected to be a significant step forward for future wireless convergence applications.

Keywords: communication and sensing; cyclic prefix interval (CPI); orthogonal frequency division multiplexing (OFDM); range extension

1. Introduction

Starting from Marconi's first transatlantic wireless transmission in 1899, wireless communication has been a crucial technology for developing today's modern lifestyle. There is a wide range of potential applications in wireless communication and sensing areas, such as cellular devices [1], wireless local area networks (WLANs) [2], vehicular communications [3], security scanner, biological diagnosis, non-destructive detection, and radar imaging [4]. From a technological perspective, a converged system is expected to provide enormous benefits in terms of both spectrum efficiency and cost-effectiveness [5–7]. In the past, different waveforms have been used independently for implementing most wireless communication, and sensing functionalities [8,9]; consequently, the systems are bulky, energy consumable, and uneconomical. In this context, a unified waveform simultaneously serving communication and sensing has gained substantial interest [10]. So far, the orthogonal frequency division multiplexing (OFDM) technique is well known for its benefits for wireless communications, and has not only been adopted in numerous standards but is also considered as a strong candidate for future wireless communication systems (5G and beyond) [11,12]. More interestingly, the OFDM waveform has also been well documented for its effectiveness in radar applications [13–15]. Therefore, OFDM waveforms are promising for the convergence of communication and sensing [16–20].

The OFDM wireless communication technically requires inverse fast Fourier transform (IFFT) and fast Fourier transform (FFT) operations to transmit and receive data. The cyclic prefix interval (CPI), also known as a guard interval, makes OFDM transmission robust against multi-path radio channel. However, under the channel impulse response longer than the CPI, inter-symbol interference (ISI) degrades communication performance, and in mobility scenarios, inter-carrier interference (ICI) causes orthogonality loss among the subcarriers and ISI as a consequence. There are some approaches to equalize this issue in communication, for instance, basis-expansion-model-based channel transformation [21], iterative finite length-equalization technique [22], and adjusting the CPI length according to the channel length [23].

1.1. Related Works

The OFDM waveform for sensing can be processed either by the conventional correlation-based approach [24,25], or by OFDM symbol-based processing [26]. Correlation-based sensing is usually performed by cross-correlation in the delay and Doppler domains between the transmitted and received pulses, and different schemes have been proposed to improve sensing performance. For example, a good approximation of the transmitted signal is generated at the receiver for removing clutter in the correlation-based target detection [15]. Work in [25] proposes to use the information of data symbols for ambiguity suppression, and circular correlation for range extension up to an OFDM symbol duration. Different correlation-based OFDM radar receiver schemes have been compared in [27], in terms of complexity, signal-to-interference-plus-noise-ratio, and robustness against ground clutter.

Alternatively, similar to OFDM-based communication, OFDM-based sensing can also use IFFT/FFT operations to extract range and speed information. Based on this approach, a 77 GHz OFDM-based sensing system with a bandwidth of 200 MHz demonstrated a sensing resolution of 0.75 m with the maximum range of 150 m [28]. Another OFDM-based radar at 77 GHz used a stepped carrier approach to achieve a sensing resolution of 0.146 m with a bandwidth of 1.024 GHz, while the maximum range is 60 m [29]. Moreover, the authors implemented OFDM-based radar processing for automotive scenario by using a relatively longer interval of 128 ms to achieve speed resolution of 0.22 m/s, while the range resolution was 1.87 m for a bandwidth of 80 MHz at 5.2 GHz [30].

These two sensing processing approaches were employed in the development of OFDM-based radars, while from the viewpoint of converging OFDM-based communication and sensing, OFDM symbol-based sensing processing is more attractive, provided that a sensing receiver is synchronized with the transmitter and the transmitted data are readily available for sensing processing. Some interesting research has been done on OFDM-based convergence in the microwave band. By using OFDM waveforms which are designed for 3GPP-LTE and 5G-NR at 2.4 GHz with a bandwidth of 98.28 MHz, OFDM-based sensing supports a sensing resolution of 1.5 m and a maximum range of 350 m and performs an algorithm for self-interference cancellation in the full-duplex mode [31]. Authors in [32] provide measurement results for the indoor mapping using a 28 GHz carrier frequency for the 5G-NR with a bandwidth of 400 MHz and achieve a sensing resolution of 0.4 m. Another work in [33] shows results of mmWave demonstration testbed for joint sensing and communication; measurements were performed at 26 GHz with a bandwidth of 10 MHz to identify the angular location of different targets using beamforming technique. The work in [34] also presents a range resolution of 1.61 m and a maximum range of 206 m within 93 MHz bandwidth at the 24 GHz band. In addition, authors in [35] provide a parameter selection criterion for joint OFDM radar and communication systems by considering vehicular communication scenarios, such as CPI, subcarriers spacing, and coherence time of the channel.

1.2. Motivation and Contribution

Please note that enabling the sensing functionality of the OFDM waveform (which is designed for wireless communication) does not provide the flexibility of parameter adjustment according to the sensing requirements. Furthermore, the ISI cancellation/compensation techniques proposed for OFDM wireless communication are not differently applicable for OFDM-based sensing because the transformation or truncation-based equalization destructs the sensing information. Ideally, the delay of an echo for sensing should fall within the CPI, and the Doppler frequency normalized over OFDM waveform interval should be an integer. However, in a real scenario for sensing, a target is located randomly and moves with an arbitrary speed. Consequently, an OFDM waveform designed for communication shows limitation in obtaining high sensing resolution and a large detection range.

As we know, the detection range of a single target is determined by the detectable OFDM signal strength and an adjustment of delay offset. In the case of multiple echoes with delay beyond the CPI, the OFDM-based sensing is mainly limited by the ISI, free-space-path-loss (FSPL), and processing gain. Echoes outside the CPI cause ISI as previous OFDM symbols interfere with current OFDM symbol in the processing window, which increases the threshold for target detection. In addition, echoes with delay longer than the CPI will achieve less processing gain, which reduces linearly with the delay. This loss of processing gain along with the ISI makes it difficult for OFDM-based sensing to detect targets outside the CPI, particularly in the millimeter-wave region featuring large bandwidth and high FSPL. Therefore, the extension of sensing range beyond the conventional limit of CPI is one of the important issues in developing communication and sensing converged systems for applications such as indoor mapping, digital health monitoring, unmanned aerial vehicles, and residential security.

In this work, we propose and experimentally demonstrate a converged communication and sensing system operating at 97 GHz using the same 16-QAM (quadrature amplitude modulation) OFDM waveform. An approach based on zero-delay shift is proposed to extend the detectable range by compensating for the IFFT processing gain for echoes outside the CPI. In the proposed method, we extended the range of an OFDM-based sensing, while the simplicity of operations for range and speed estimation is achieved using IFFT/FFT operations. The proposed method uses delay-shifts in the received signal before processing a received OFDM symbol. Active subcarriers in the received OFDM symbol are divided by the active subcarriers in the current and previous transmitted OFDM symbols (employed number of transmitted OFDM symbols determine the rang extension), and IFFT operations are used after each delay-shift to generate matrices in the delay and delay-shift domains (delay domain is the result of IFFT operation). Delay-shift rows at delay zero are concatenated to extend the delay-shift domain. Concatenation of delay-shift rows for different received OFDM symbols provides a matrix in delay-shift and time domain, and FFT operations over time domain provide the speed estimation. An experiment with a heterodyne W-band transmitter/receiver is performed, and both sensing and communication performance are measured in terms of range/speed profile and bit-error-rate (BER). The proposed approach for range extension is verified for distances well beyond the CPI and provides a range resolution of 0.042 m, and speed resolution of 0.79 m/s using a single OFDM waveform, which is promising in driving OFDM-based converged systems for future applications.

The rest of this paper is organized as follows. Section 2 presents the model for the OFDM-based converged system to provide the details of extracting sensing information from the received OFDM waveform. Section 3 details the proposed method for range extension in an OFDM-based converged system. Section 4 provides simulation results, while Section 5 is dedicated to experimental measurement results and discussions. Section 6 provides the conclusion of this work.

2. Communication and Sensing Convergence Using OFDM Waveforms

Motivated by the OFDM-based sensing presented in [26,34], a reference system model for the convergence of communication and sensing is presented here. An OFDM waveform for communication purposes consists of several OFDM symbols, each with orthogonal subcarriers modulated by data symbols and cyclically extended by appending the last part of the signal at the beginning called cyclic prefix (CP). If Δf represents the subcarriers spacing, N the number of orthogonal subcarriers, T the OFDM symbol duration, T_{cp} as the CPI, $T_s = T_{cp} + T$ the effective duration of the OFDM symbol, and M the number of OFDM symbols, then the analytical expression of the transmitted OFDM waveform is [34],

$$s(t) = \sum_{\mu=0}^{M-1} \sum_{n=0}^{N-1} S(\mu N + n) e^{(j2\pi n \Delta f(t - \mu T_s))} e^{(-j2\pi n \Delta f T_{cp})} \text{rect}\left(\frac{t - \mu T_s}{T_s}\right), \quad (1)$$

where $S(\mu N + n)$ is the data symbol at nth subcarrier of μth OFDM symbol. The rect(t) in (1) is the rectangular pulse shape, such that rect$(t) = 1$ for $t \in [0 \ 1]$ and 0 otherwise. The term $\exp(-j2\pi n \Delta f T_{cp})$ appears due to the cyclic extension of OFDM symbols by the CP.

In order to fulfill the orthogonality among subcarriers, over the interval T, the following condition must be held:

$$\Delta f = \frac{1}{T}, \quad (2)$$

and T_{cp} should accommodate the maximum expected delay caused by the radio channel. The baseband signal $s(t)$ is up-converted by a carrier frequency f_c to form $\tilde{s}(t)$ for transmission,

$$\tilde{s}(t) = s(t) e^{j2\pi f_c t}. \quad (3)$$

The received signal $\tilde{a}(t)$ at the sensing receiver is the sum of echoes from different targets. Using point-target channel model for L number of targets,

$$\tilde{a}(t) = \sum_{l=1}^{L} b_l \tilde{s}(t - \tau_l), \quad (4)$$

where τ_l and b_l represent delay and attenuation related to the lth target, respectively. If lth target is located at a distance R_l and moving with a speed of v_l, delay τ_l in the received echo can be expressed as

$$\tau_l = \frac{2(R_l - v_l t)}{c_0}, \quad (5)$$

and b_l [34],

$$b_l = \sqrt{\frac{c_0^2 G_{Tx} G_{Rx} \sigma_{RCS_l}}{(4\pi)^3 R_l^4 f_c^2}}, \quad (6)$$

where in (6), c_0 is the speed of light in free space; σ_{RCS_l} is the radar cross-section of the lth target; and G_{Tx}, G_{Rx} represent transmitting and receiving antenna gain, respectively. For the communication link, the signal attenuation b_{com} is

$$b_{com} = \sqrt{\frac{c_0^2 G_{Tx} G_{Rx}}{(4\pi)^2 R_{com}^2 f_c^2}}, \quad (7)$$

where R_{com} indicates the distance of the communication link.

For sensing processing, a single target is sufficient for mathematical derivations due to the linear operation in (1). The analytical expression for the received echo from a target, located at a distance R, moving with the speed of v, and attenuated by $\hat{b}$ (assuming constant attenuation factor for frequencies within the bandwidth) is obtained by using delay $\tau = (2R - 2vt)/c_0$ in (1), i.e.,

$$\tilde{a}(t) = \sum_{\mu=0}^{M-1} \sum_{n=0}^{N-1} \hat{b} S(\mu N + n) e^{\left(j2\pi n \Delta f (t - \frac{(2R-2vt)}{c_0}) - \mu T_s - T_{cp}\right)}$$

$$\cdot e^{\left(j2\pi f_c (t - \frac{(2R-2vt)}{c_0})\right)} \text{rect}\left(\frac{t - \left(\frac{2R-2vt}{c_0}\right) - \mu T_s}{T_s}\right) + \hat{z}(t), \quad (8)$$

where $\hat{z}(t)$ is to account for the additive white Gaussian noise (AWGN). Since f_c is usually very high compared to the bandwidth of the signal, in particular in the millimeter-wave band, the Doppler shift $(n\Delta f 2v)/c_0$ is negligible for the subcarriers and the overall Doppler shift appears only caused by $(2vf_c)/c_0$.

The received signal is down-converted to baseband, which is equivalent to

$$a(t) = \sum_{\mu=0}^{M-1} \sum_{n=0}^{N-1} b S(\mu N + n) e^{\left(j2\pi n \Delta f (t - \frac{(2R)}{c_0} - \mu T_s - T_{cp})\right)} e^{\left(j2\pi \frac{2vf_c}{c_0} t\right)} \text{rect}\left(\frac{t - (\frac{2R}{c_0}) - \mu T_s}{T_s}\right) + z(t), \tag{9}$$

where b and $z(t)$ represent $\hat{b}\exp(-j2\pi f_c 2R/c_0)$ and $\hat{z}(t)\exp(-j2\pi f_c t)$, respectively.

Finally, the signal $a(t)$ is sampled, the CP part is removed before it is converted into the frequency domain by using FFT operation,

$$A(\mu N + n) = b S(\mu N + n) e^{\left(-j2\pi n \Delta f \frac{2R}{c_0}\right)} e^{\left(j2\pi \frac{2vf_c}{c_0} \mu T_s\right)} + Z(\mu N + n), \tag{10}$$
$$n \in [0, N-1], \quad \mu \in [0, M-1]$$

where $A(\mu N + n)$ and $Z(\mu N + n)$ are frequency domain equivalents of $a(t)$ and $z(t)$.

Once the received signal is translated back into the frequency domain, the element-wise division of the received OFDM symbol by the respective transmitted OFDM symbol is performed to construct the channel matrix $\mathbf{H}$, i.e.,

$$H(\mu N + n) = b e^{\left(-j2\pi n \Delta f \frac{2R}{c_0}\right)} e^{\left(j2\pi \frac{2vf_c}{c_0} \mu T_s\right)} + \frac{Z(\mu N + n)}{S(\mu N + n)}, \tag{11}$$

where $H(\mu N + n)$ represents the μth column and nth row of the channel matrix $\mathbf{H}$, and $Z(\mu N + n)/S(\mu N + n)$ defines the noise floor that depends on the digital modulation, e.g., a 16-QAM mapping affects the noise floor by approx 2.7 dB [36]. The IFFT of $\mathbf{H}$, along subcarriers, provides the range information,

$$r(d) = e^{\left(j2\pi \frac{2vf_c}{c_0} \mu T_s\right)} \frac{b}{N} \sum_{n=0}^{N-1} e^{\left(-j2\pi n \Delta f \frac{2R}{c_0}\right)} e^{\left(j\frac{2\pi}{N} nd\right)} + \check{z}(d), \tag{12}$$
$$d \in [0, N-1]$$

where $\check{z}(d)$ represents the noise part.

The $|r(d)|$ shows a peak value under the following condition:

$$d = \left\lfloor \frac{2R \Delta f N}{c_0} \right\rfloor, \tag{13}$$

i.e., the value of d corresponding to the maximum of $|r(d)|$ holds the information of the target range, and the range resolution ΔR (minimum distinguishable distance between the two targets) is defined as

$$\Delta R = \frac{c_0}{2\Delta f N}. \tag{14}$$

Similarly, the FFT operation over different OFDM symbols in $\mathbf{H}$ (over μ domain in (11)) provides the information about the speed of the target and can be recognized by using

$$p = \left\lfloor \frac{2vf_c T_s M}{c_0} \right\rfloor, \quad p \in [0, M-1] \tag{15}$$

whereas the speed resolution Δv can be calculated by setting $p = 1$,

$$\Delta v = \frac{c_0}{2f_c T_s M}. \tag{16}$$

The IFFT/FFT operations on **H** provide processing gain due to coherent addition of signals, and the overall processing gain is

$$G = NM. \tag{17}$$

If there are $\acute{N}$ guardband subcarriers on each side of the OFDM symbol, then the processing gain reduces to $(N - 2\acute{N})M$. Here, it is important to note that the guardband subcarriers reduce the bandwidth of the OFDM waveform, and consequently the value of ΔR increases in (14) when N is replaced by $(N - 2\acute{N})$. Although ΔR increases due to guardband subcarriers, improvement in the range accuracy (resolution of IFFT) is linked to the size of IFFT [37].

Figure 1 highlights the sensing processing using **H**. Figure 1a shows the real part of the channel matrix **H**, which has sinusoidal variations due to the range and speed of a single target. The IFFT operation, along with subcarriers, identifies the delay associated with the range, as shown in Figure 1b. Afterwards, the FFT operation provides the sensing information in the delay-Doppler profile, as shown in Figure 1c,d.

It is clear that the sensing performance depends on the OFDM waveform parameters because bandwidth defines the range resolution, and the duration of the OFDM waveform determines speed resolution. For the speed, the upper limit is selected as $\Delta f > 20 f_c v/c_0$ to maintain the acceptable level of orthogonality among the subcarriers [35].

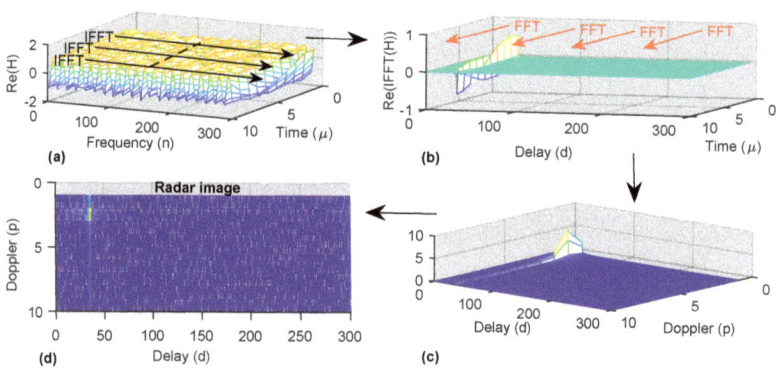

Figure 1. OFDM–based sensing from the channel matrix **H**. (**a**) IFFT operation over subcarriers. (**b**) FFT operation over OFDM symbols. (**c**) 3-D plot of the delay–Doppler profile. (**d**) Radar image, indicating the delay associated to the range and the Doppler frequency related to the speed of the target.

3. Proposed Method for Range Extension

In OFDM-based sensing, the maximum range is limited by the CPI [28,35] as echoes falling outside the CPI cause ISI and suffer in processing gain. We propose a zero-delay shift method to compensate for delay τ in an echo to maintain its processing gain G during IFFT operation for OFDM-based sensing.

Using sampling intervals $\Delta T = 1/F_s$ and $\Delta f = 1/(N \Delta T)$, the sampled version of (9) is

$$a(k\Delta T) = \sum_{\mu=0}^{M-1} \sum_{n=0}^{N-1} bS(\mu N + n) e^{\left(j2\pi n \frac{1}{(N\Delta T)} \left(k\Delta T - \frac{2R}{c_0 \Delta T} - \mu \frac{T_s}{\Delta T} - \frac{T_{cp}}{\Delta T} \right) \right)}$$
$$\cdot e^{\left(j2\pi \frac{2vf_c}{c_0} k\Delta T \right)} \text{rect}\left(\frac{k\Delta T - \frac{2R}{c_0} - \mu \frac{T_s}{\Delta T}}{\frac{T_s}{\Delta T}} \right) + z(k\Delta T), \tag{18}$$

where k is the sampled time index. Using $F_s/\Delta f = T/\Delta T = N$, $T_s/\Delta T = N_s$, $T_{cp}/\Delta T = N_{cp}$, $m = \lfloor (2R)/(c_0 \Delta T) \rfloor$, and $a(k)$, $z(k)$ to represent $a(k\Delta T)$ and $z(k\Delta T)$ respectively,

$$a(k) = \sum_{\mu=0}^{M-1} \sum_{n=0}^{N-1} bS(\mu N + n) e^{\left(j2\pi \frac{n}{N}(k - m - \mu N_s - N_{cp}) \right)}$$

$$\cdot e^{\left(j2\pi \frac{2v f_c}{c} k \Delta T \right)} \text{rect} \left(\frac{k - m - \mu N_s}{N_s} \right) + z(k). \quad (19)$$

$$k \in [0, MN_s - 1]$$

A delay-shift in k by m samples shifts the target at zero on the delay axis. Since m is unknown, sequentially increasing the delay-shift in k identifies m when a peak appears at delay zero. This process can identify echoes with delay longer than the CP, provided they arrive with detectable signal strength. If we extend the sensing range up to Q number of OFDM symbols, the proposed method can be described in following steps:

1. N samples are selected (window of length N) from the received OFDM waveform to perform an N-point FFT operation;
2. Received data symbols are divided by the transmitted data symbols in the current and previous $Q - 1$ OFDM symbols;
3. N-point IFFT operation is performed on results obtained in step 2, which provides first columns (in the delay domain) for Q number of matrices;
4. Steps 1–3 are repeated for N_s number of delay-shifts in the selected window. Completion of this step provides Q number of matrices each of size $N \times N_s$ in delay and delay-shift domains;
5. Delay-zero rows ($d = 0$) of each of the Q number of matrices, generated in step 4, are concatenated to form a row of another matrix, which will be used for range/speed processing;
6. At this point, the selected window of N samples has been shifted by N_s samples; now, OFDM symbols, which are used for division in step 2, are replaced by next OFDM symbol, e.g., $[S_q, S_{q-1}, \ldots, S_{q-(Q-1)}]$ are replaced by $[S_{q+1}, S_q, \ldots, S_{q-(Q-2)}]$, where S_q denotes qth OFDM symbol;
7. Steps 1–6 are repeated M times.

Using the above process, a matrix (in the delay-shift and time domains) of size $QN_s \times M$ is constructed, which requires an M-point FFT operation to complete the range/speed plot.

Figure 2 shows different steps in the proposed method to detect two targets separated by more than one OFDM symbol duration; $Q = 2$ is used, and the current OFDM symbol number is q. Figure 2a,b are obtained using steps 1–4 of the proposed method; Figure 2c is the plot of step 5 and using $M = 12$ for step 6; Figure 2d is the final range/speed plot, where the range is extended up to two OFDM symbols. A schematic diagram of the proposed method is presented in Figure 3. Delay shifts are used in the sampled version of the incoming OFDM waveform to get the frequency domain signal Y_q. Element-wise division of Y_q is performed with the current OFDM symbol S_q and previous OFDM symbols for sensing and combining.

We use periodogram to compare the performance of the proposed method with conventional OFDM-based sensing. Periodogram of the conventional OFDM-based sensing is defined as [31],

$$D_{\text{conv}(d,p)} = \left| \sum_{\mu=0}^{M-1} r(d) e^{\frac{-j2\pi \mu p}{M}} \right|^2, \quad (20)$$

$$d \in [0, N-1] \quad p \in [0, M-1]$$

where $r(d)$ is defined in (12). A target is detected if the peak in $D_{\text{conv}(d,p)}$ is above a threshold level (usually defined by the minimum detectable signal strength). For the range/speed plot, $\mathbf{D}$ is often transformed to normalized power, and in dB scale using $10\log_{10}(\mathbf{D}/max[\mathbf{D}])$, where $max[\mathbf{D}]$ represents the maximum value of $\mathbf{D}$.

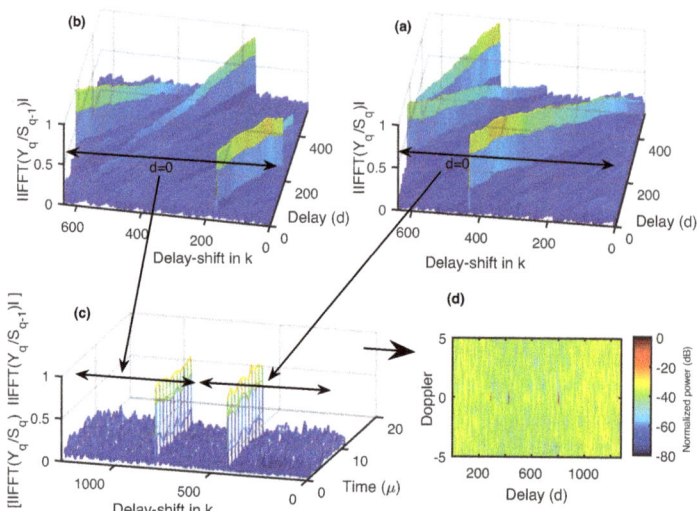

Figure 2. Explanation of the proposed method when range extension is up to two OFDM symbols. (**a**,**b**) Steps 1–4 of the proposed method provide two matrices. (**c**) Delay–zero rows of the matrices in (**a**,**b**) are concatenated according to the step 5. (**d**) Range/speed plot is completed using M–point FFT operation over time domain.

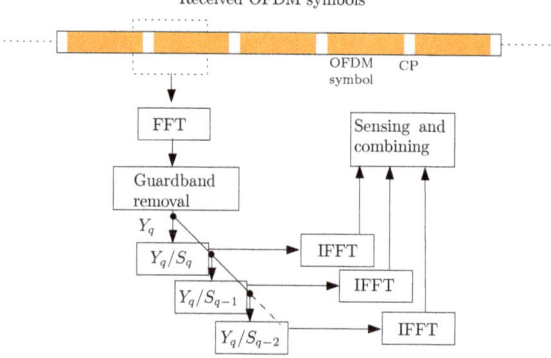

Figure 3. Schematic of the proposed range extension method to detect targets beyond the range limit in the conventional OFDM–based sensing.

Similarly, the periodogram of the proposed method is

$$D_{\text{pro}(\hat{d},p)} = \left| \sum_{\mu=0}^{M-1} \hat{r}_{\hat{d}}(0,\mu) e^{\frac{-j2\pi\mu p}{M}} \right|^2, \quad \hat{d} \in [0, (Q-1)N_s - 1] \quad p \in [0, M-1] \quad (21)$$

where $\hat{d}$ represents the delay-shift domain and $\hat{r}_{\hat{d}}(0,\mu)$ is obtained by concatenation of Q segments as defined in step 5 of the proposed method, i.e.,

$$\hat{r}_{\hat{d}}(0,\mu) = [r_{\hat{d}}(0,\mu), r_{\hat{d}}(0,\mu-1), ..., r_{\hat{d}}(0,\mu-(Q-1))], \quad \hat{d} \in [0, N_s - 1] \quad (22)$$

where $r_{\acute{d}}(0,\mu)$ is obtained by using $\acute{d}=0$ in $r_{\acute{d}}(d,\mu)$,

$$r_{\acute{d}}(d,\mu) = \frac{1}{N}\sum_{n=0}^{N-1} H_{\acute{d}}(\mu N+n)e^{\frac{j2\pi nd}{N}}, \quad d \in [0, N_s-1] \quad (23)$$

i.e.,

$$r_{\acute{d}}(0,\mu) = \frac{1}{N}\sum_{n=0}^{N-1} H_{\acute{d}}(\mu N+n). \quad (24)$$

If we represent $m = gN_s + \hat{m}$ where $g \in [0, Q-2]$ and $\hat{m} \in [0, N_{cp}-1]$, received qth OFDM symbol and transmitted OFDM symbols provide

$$\mathbf{H}_{\acute{d},(q-i)} = \frac{\mathbf{Y}_{\acute{d},q}}{\mathbf{S}_{(q-i)}}, \quad (25)$$

where $\mathbf{H}_{\acute{d},(q-i)}$ is taken as simplified notation for $H_{\acute{d}}((q-i)N+n)$, $i \in [0, Q-1]$, and $\mathbf{Y}_{\acute{d},q} = \text{FFT}(A_{\acute{d}}(k))$, where $A_{\acute{d}}(k)$ is the delay-shift of $A(k)$ by $\acute{d}$ and $k \in [\acute{d}+(q)N_s - gN_s - \hat{m}+, \acute{d}+(q+1)N_s - gN_s - \hat{m} - N_{cp}]$ (interval of the N-samples of the waveform is selected at the initial step of the proposed method). It is clear that at $\acute{d} = \acute{d}_0 = \hat{m} + N_{cp}$, $A_{\acute{d}_0}(k)$ represents $(q-g)$th OFDM symbol without N_{cp}; hence, (25) changes to

$$\mathbf{H}_{\acute{d}_0,(q-i)} = \frac{b\mathbf{S}_{(q-g)} + \mathbf{Z}_{(q-g)}}{\mathbf{S}_{(q-i)}}, \quad (26)$$

where $\mathbf{Z}_{(q-g)}$ represents noise part in the $(q-g)$th OFDM symbol. Similarly, at $\acute{d} = \acute{d}_1 = (\hat{m}+N)$, $A_{\acute{d}}(k)$ consists of last N_{cp} samples of the $(q-g)$th OFDM symbol and $N-N_{cp}$ samples of $(q-g+1)$th OFDM symbol; therefore,

$$\mathbf{H}_{\acute{d}_1,(q-i)} = \frac{N-N_{cp}}{N\mathbf{S}_{(q-i)}}b\mathbf{S}_{(q-g)}e^{-\frac{j2\pi n}{N}N_{cp}} + \frac{N_{cp}}{N\mathbf{S}_{(q-i)}}b\mathbf{S}_{(q-g+1)} + \frac{\mathbf{Z}_{(q-g)}}{\mathbf{S}_{(q-i)}}. \quad (27)$$

Using (26) in (24) provides the maximum of $r_{\acute{d}}(0,\mu)$, which is same as $r(d)$ in (20), whereas (27) indicates the additional peak with height reduced by a factor of N_{cp}/N and affected by the ISI. Similar to $D_{\text{conv}(d,p)}$, where a processing of NM is assigned to a peak, $D_{\text{pro}(\hat{d},p)}$ also provides the same processing gain when $i = g$ in (26); otherwise, $\mathbf{H}_{\acute{d},(q-i)}$ in (26) is interference. In (27), contrary to interference term, N_{cp} samples are coherently added when used in (24) and the processing gain is $10\log_{10}(N_{cp}^2/(N-N_{cp}))$.

In a generalized scenario, there can be L echoes with delays not limited to CPI; the received signal $y(k)$ is the summation of all echoes, each represented by the (19),

$$y(k) = \sum_{l=0}^{L-1}\sum_{\mu=0}^{M-1}\sum_{n=0}^{N-1} b_l S(\mu N+n)e^{\left(j2\pi\frac{n}{N}(k-m_l-\mu N_s-N_{cp})\right)} \\ \cdot e^{\left(j2\pi\frac{2v_lf_c}{c}k\Delta T\right)}\text{rect}\left(\frac{k-m_l-\mu N_s}{N_s}\right) + z(k), \quad (28)$$

where m_l represents the delay associated with lth echo. Based on the delay m_l, we split the $y(k)$ into three portions such as $y_1(k)$ for $m_l \leq N_{cp}$, $y_2(k)$ for $N_{cp} < m_l \leq N_s$, and $y_3(k)$ for $m_l > N_s$, i.e.,

$$y(k) = y_1(k) + y_2(k) + y_3(k) + z(k). \quad (29)$$

Since y_3 is formed by the summation of echoes that are outside the current OFDM symbol, therefore this part is only ISI. Unlike $y_3(k)$, the ISI part of $y_2(k)$ increases as m_l approaches to N_s. The detection of the echoes in $y_2(k)$ and $y_3(k)$ is possible if the processing gain G is sufficient to overcome the related ISI and noise.

3.1. Signal-to-Interference Ratio

In the proposed method, the signal-to-interference ratio (SIR) changes with the shifting of k. At the sensing receiver, strength of an echo depends on several factors such as antenna gain, round trip distance from the target, carrier frequency, and radar cross-section, as mentioned in (6). If we define P_{Rx_1} as the received power of $y_1(k)$, P_{Rx_2} for $y_2(k)$, and P_{Rx_3} for $y_3(k)$, then the SIR during the shifting of k is as below.

- At the beginning, echoes within $y_1(k)$ are detected under the collective ISI caused by $y_3(k)$ and $y_2(k)$, and we can define the SIR during this process as

$$\text{SIR}_1 = \frac{P_{Rx_1}}{\alpha_2(\hat{k})P_{Rx_2} + P_{Rx_3}}, \tag{30}$$

where $\hat{k}$ indicates the shift in k and $0 \leq \alpha_2(\hat{k}) < 1$ defines the part of P_{Rx_2} appearing as interference. $\alpha_2(\hat{k}) = 0$ indicates that there are no echoes to form $y_2(k)$.

- At the second stage, when $\hat{k}$ is beyond the N_{cp} and within N_s, echoes that form $y_2(k)$ are detected and a part of $y_1(k)$ causes ISI, which increases with $\hat{k}$. The SIR can be defined as

$$\text{SIR}_2 = \frac{P_{Rx_2}}{\alpha_1(\hat{k})P_{Rx_1} + P_{Rx_3}}, \tag{31}$$

where $0 \leq \alpha_1(\hat{k}) < 1$ is used to account for the ISI caused by part of P_{Rx_1}.

- Similarly, when we detect echoes in $y_3(k)$, the SIR is

$$\text{SIR}_3 = \frac{P_{Rx_3}}{P_{Rx_1} + \alpha_2(\hat{k})P_{Rx_2}}, \tag{32}$$

where P_{Rx_1} appears as ISI because at this stage the element-wise division is performed by the previous OFDM symbol S_{q-1}.

Here, it is important to mention that $P_{Rx_1} > P_{Rx_2} > P_{Rx_3}$ (assuming same radar cross-section for different targets associated with echoes) because of the FSPL difference between echoes that form $y_1(k)$, $y_2(k)$ and $y_3(k)$. Therefore, $\text{SIR}_1 > \text{SIR}_2 > \text{SIR}_3$, which clearly indicates that detection of echoes in $y_2(k)$ and $y_3(k)$ is not possible without the sufficient processing gain obtained through the IFFT/FFT operation during sensing. Usually, the Doppler estimation requires large interval (compared to the OFDM symbol duration) of the waveform; therefore, large number of OFDM symbols can provide sufficient processing gain for the echoes to overcome ISI.

3.2. Effect of CP

For $y_3(k)$, during delay-shifting stage of the proposed method, the last part of the window, which is selected in step 1 of the proposed method, occupies the complete CP part of the OFDM symbol, and a peak with processing gain of $20\log_{10}(N_{cp}) - 10\log_{10}(N - N_{cp}) - \text{Mod}_{noise}$ dB appears at delay zero (defined by (27)). Where Mod_{noise} dB is the raise in noise due to digital modulation, e.g., 16-QAM causes a raise of ≈ 2.7 [36]. For M number of OFDM symbols, an additional $10\log_{10}(M)$ dB is added to CP peak. Appearance of peaks due to CP, for echoes with delay longer than OFDM symbol duration, which are exactly N samples behind the target, can be eliminated from the observations.

3.3. Computational Complexity of the Proposed Method

The computational complexity is measured in terms of number of complex multiplications and additions. It is considered that removal of the CPI is negligible in complexity, divisions are equivalent to multiplications, and performing an IFFT/FFT of size N requires $(N/2)\log_2(N)$ number of complex multiplications and $N\log_2(N)$ number of complex additions [27]. Table 1 provides the complexity of the proposed method and the conventional OFDM-based sensing. The complexity of the proposed method is higher by a factor of

$\approx QN_s$ because whole chain of operation, for range detection, is performed after each shift in k with maximum shifts as QN_s. For speed, N number of M-point FFTs is increased to QN_s.

Table 1. Computational complexity of the proposed method.

OFDM-Based Sensing			Proposed Method		
Operation	Complex Multiplications	Complex Additions	Operation for QN_s Delays	Complex Multiplications	Complex Additions
Division to get **H**	$4MN_{ac}$	0	Division to get **H**	$4MN_{ac}QN_s$	0
N-point IFFT	$M(N/2)\log_2(N)$	$MN\log_2(N)$	N-point IFFT	$(QN_s)M(N/2)\log_2(N)$	$(QN_s)MN\log_2(N)$
M-point FFT	$N(M/2)\log_2(M)$	$MN\log_2(M)$	M-point FFT	$(QN_s)(M/2)\log_2(M)$	$(QN_s)M\log_2(M)$
Total	$(MN/2)\log_2(MN)$	$MN\log_2(MN)$	Total	$(QN_s)(M/2)(N\log_2(N)+\log_2(M))$	$(QN_s)M(N\log_2(N)+\log_2(M))$

4. Simulation Results

To verify the processing of the proposed OFDM-based converged system, a baseband equivalent model is implemented and simulated in MATLAB. An impulse response, having taps at the round-trip delay of the targets, and each tap varying over OFDM symbols according to the complex exponential of the Doppler frequency, is used to represent the sensing channel, whereas SNR = 15 dB is set for simulation. Equal signal strength is used for different targets, while other parameters, listed in Table 2, are selected to match parameters used in our experiment.

Table 2. Simulation parameters.

Sampling frequency F_s	6 GHz
Carrier frequency f_c	97 GHz
FFT size N	512
Subcarrier spacing Δf	11.71875 MHz
No. of active subcarriers N_{ac}	300
Bandwidth B	3.9 GHz
Bandwidth occupied by active subcarriers $N_{ac}\Delta f$	3.51 GHz
Bandwidth utilization	90%
Digital mapping	16 QAM
Effective OFDM symbol duration T_s	0.1066 μs
Cyclic prefix duration (CP)	128 samples, 0.0213 μs
No. of OFDM symbols M	36 for range 18,432 for speed
Range resolution $\Delta R = c/(2\Delta f N_{ac})$	0.042 m
Unambiguous range $(N-1)c/(2\Delta f N)$	12.775 m
Maximum range (within CPI)	3.2 m
Speed resolution Δv	0.79 m/s
Unambiguous speed $\pm \frac{(M-1)\Delta v}{2}$	±7464 m/s

It is important to note that for scenarios where we use only static targets, $M = 36$ is used, which provides sufficient gain for range detection but results in a high value of Δv in the range/speed plot, although this is irrelevant for static targets. For the range/speed plot, the absolute of the delay/Doppler matrix is first normalized to unit (by dividing with the maximum absolute value) and then converted to dB scale. Figure 4a–c presents the range plots for static targets at distances 0.6 m, 1.3 m 1.5 m, and 10 m, respectively, using the conventional OFDM-based sensing. Results show that the targets are identified correctly at the distances in Figure 4a,b.

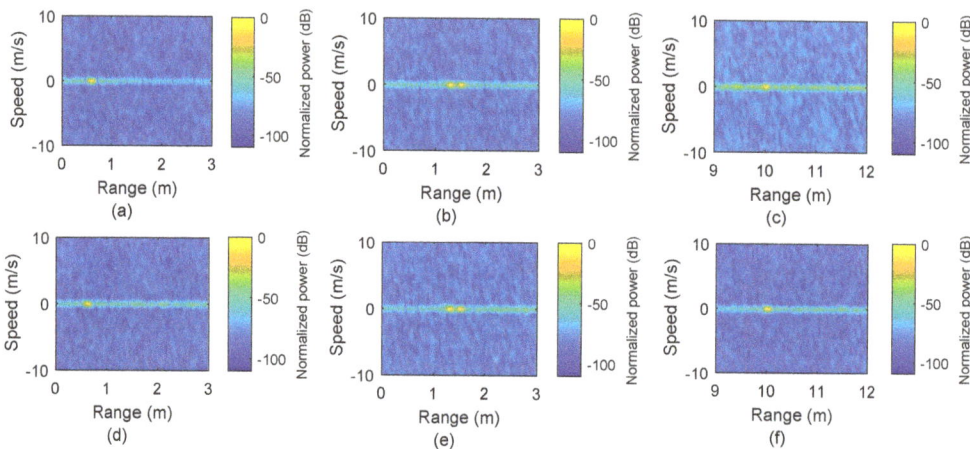

Figure 4. Simulation results of the conventional OFDM–based sensing vs. the proposed range extension method. (**a–c**) Range plots for a target at 0.6 m, two targets at 1.3 m and 1.5 m, and a target at 10 m, respectively, by using conventional OFDM–based sensing. (**d–f**) Range plots obtained by using the proposed range extension method.

However, Figure 4c clearly shows that the sensing performance has been compromised, in terms of SNR, for the target located at a 10 m distance. This reduction of SNR happens because the CPI covers a range up to 3.2 m. Beyond this range limit, ISI occurs, and the processing gain is also reduced for that target located at 10 m. In comparison, the scenario mentioned above is also processed for sensing using our proposed method by finding the zero-delay for each target, as shown in Figure 4d–f. Our proposed method offers better performance for the target at 10 m by avoiding the loss in processing gain.

Figure 5 represents the results when multiple targets exist and one of them is moving. In Figure 5a, the conventional approach provides accurate results for two targets, one static target located at 0.6 m and the other target moving with a speed of 5 m/s and situated at 0.85 m. In comparison to Figure 5a, Figure 5b represents the results obtained by using our proposed method where peak height is similar to the peak height in Figure 5a, but ISI effect is used for locations away from the targets.

In order to verify the proposed method for the range beyond the OFDM symbol, we also simulated the case that three targets are placed at 0.85 m (moving with 2.34 m/s), 12.65 m, and 17.65 m, respectively, and the sensing results are shown in Figure 5c,d. The farthest target at 17.65 m is beyond the range of an OFDM symbol duration (12.77 m) and it does not appear in Figure 5c using the conventional approach, but it is detectable with our proposed method, as shown in Figure 5d. There is another peak with height ≈ -20 dB at 4.88 m (12.77 m behind the target at 17.65 m) in Figure 5d, which appears due to the CP effect.

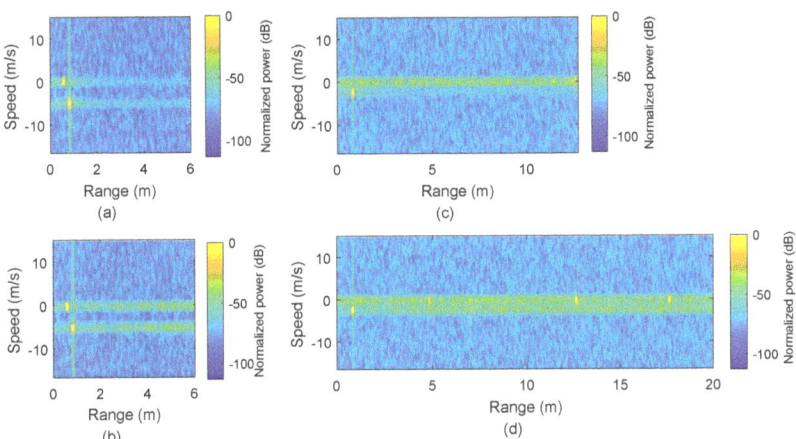

Figure 5. (**a**) One static target at 0.6 m and one moving with 5 m/s using the conventional OFDM–based sensing. (**b**) Results by using the proposed range extension method for the scenario in (**a**). (**c**) Range–speed plot for targets at 0.85 m and moving with 2.34 m/s, with second and third static targets at 12.65 m and 17.65 m, respectively. (**d**) Range–speed plots obtained by using the proposed range extension method for targets mentioned in (**c**).

5. Experimental Setup and Results

In this work, we also implement an experimental demonstration. Figure 6 shows the configuration of our system in the experiment, which is composed of several blocks in the digital and analog domain for transmission and reception. In our experiment, the baseband 16-QAM OFDM waveform is digitally generated for communication and sensing using parameters listed in Table 2. An oversampling factor of 20 is used before the signal is digitally up-converted to an intermediate frequency (IF) at 3 GHz through the IQ mixing. The IF signal is then fed to an arbitrary waveform generator (AWG) operating at 120 GS/s. Before the free-space transmission, the signal is up-converted to the W-band with a carrier frequency of 97 GHz using a commercially available W-band mixer at 94 GHz. Subsequently, a W-band amplifier with 10 dB gain is used to boost the signal to around 0 dBm, and a pair of conical horn antennas with a gain of 30 dBi is used for transmission and reception. At the receiver, the signal is first down-converted into the IF domain using a similar W-band mixer, sampled using a digital sampling oscilloscope (DSO) (KEYSIGHT MXR608A, sampling rate of 16 GS/s, bandwidth of 6 GHz) and then processed digitally to retrieve the baseband signal for further processing. In the digital domain, typical Fourier sidelobes are suppressed by using the Hamming window.

The photos of the experimental setup are shown in Figure 7a–c. Figure 7a shows the arrangement for realizing the reflective sensing to measure the range of two static targets with flat reflective surfaces. Figure 7b shows the setup for speed and range measurement with a static target and a moving target. The speed is measured via a reflective target mounted on a belt, which moves the target along the LOS and away from the receiver with adjustable speed. In this case, the Doppler frequency shift is induced in the signal reflected from the moving target, and the down-conversion from W-band carrier frequency to an IF yields a sinusoidal of the Doppler frequency, which can be observed at the DSO. Observation of the sinusoidal-like variations in the received signals at the DSO indicates the correct measurement arrangement for speed measurement.

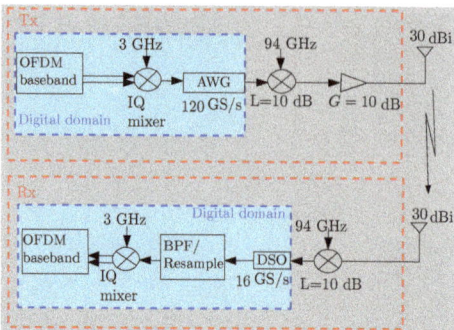

Figure 6. Block diagram of the measurement setup, showing different stages of processes in the digital and analog domains at the transmitter T_X and the receiver R_X.

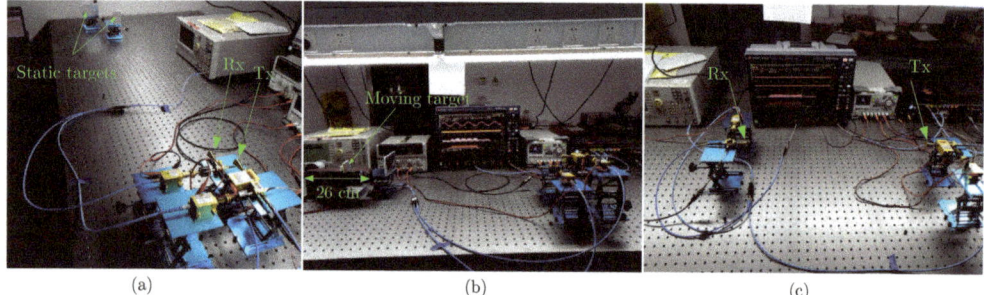

Figure 7. Experimental setup for the measurements. (**a**) Photo of the setup to measure range of two static targets. (**b**) Range-speed measurement setup of a static target and a target mounted on a belt capable to move with adjustable speed up to 5 m/s. (**c**) Data link arrangement.

The setup is calibrated during range measurement, and we measure different cases for illustrating the range and speed measurements, as listed in Table 3, where d_1 and d_2 are the actual distances of targets from the receiving antenna and $\hat{d}_1$, $\hat{d}_2$ represent measurement results. Similarly, v and $\hat{v}$ in Table 3 represent actual and measured speeds, respectively. For the data transmission part, a LOS link is established by placing the receiver at the location of the static target, as shown in Figure 7c. In the following subsections, we discuss experimental results obtained from the measurements.

5.1. Range Measurement

In Figure 8a, range measurement is implemented when a static target is placed 0.605 m from the receiver. The target is detected with an error of 0.005 m, as mentioned in Table 3. Based on Figure 8a, we can also observe multiple reflections originating from the target and wall; TS1 and TS2 indicate first and second reflections from the target, whereas W is the wall's reflection. Here, the second reflection refers to a signal that is reflected twice from the target and eventually detected by the receiver since the receiver reflects a portion of strong signal. Then, we perform the measurements for two targets placed at 0.6 m and 0.65 m, and Figure 8b shows that the closely placed targets are distinguishable from each other at their locations. Similarly, Figure 8c provides the measurement results for two targets standing at 1.313 m and 1.548 m, respectively. The change in the received power is due to a higher FSPL.

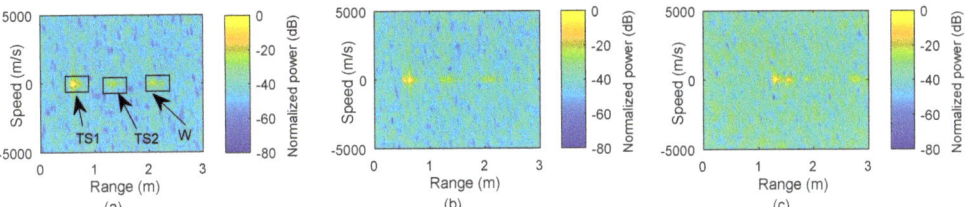

Figure 8. Measurement results for different arrangements of static targets. (**a**) Range plot of a target when placed at 0.6 m. (**b**) Range plot for two targets placed at 0.65 m and at 0.6 m. (**c**) Results of two targets placed at 1.313 m and at 1.548 m, respectively.

5.2. Range Extension

In this case, our proposed range extension method is applied using digital delay offset in the received signal to realize larger range values such as 10 m. In Figure 9, ranging performance using the conventional (Figure 9a–c) and the proposed method (Figure 9d–f) is shown, when the target location is within the CPI, within the OFDM symbol duration, and beyond the OFDM symbol duration.

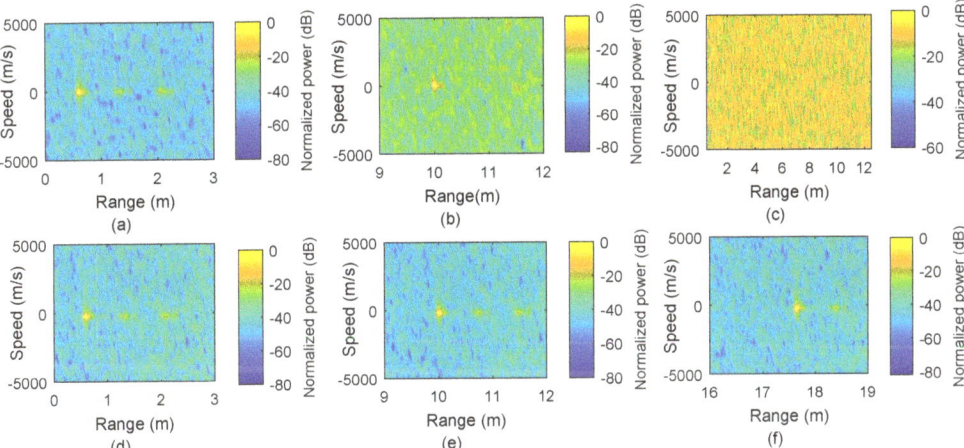

Figure 9. Range extension of the measurement results, where the range is extended through offset–delay. (**a–c**) Results of conventional approach for targets at 0.6 m, 10 m, and 17.65 m, respectively. (**d–f**) Results obtained through the proposed range extension method for the arrangements as for results in (**a–c**).

Figure 9a is the range plot for the target placed at 0.6 m in the measurement, whereas Figure 9b shows the same target shifted to 10 m, which reduces processing gain and, in turn, performance degradation. Similarly, the target in Figure 9c is beyond the OFDM symbol duration and does not appear in the conventional OFDM-based processing. As a benefit, the results obtained through our proposed method are shown in Figure 9d–f, and all targets are identifiable for all the same cases. Multiple reflections from the target and reflection from the wall appear as mentioned in Figure 8a.

Table 3. Number of targets and their arrangements for different scenarios.

Scenario	Actual Values				Measured Values			
	d_1 m	v_1 m/s	d_2 m	v_2 m/s	$\hat{d}_1$ m	$\hat{v}_1$ m/s	$\hat{d}_2$ m	$\hat{v}_2$ m/s
Single static target	0.605	0	N.A	N.A	0.60	0	N.A	N.A
Two static targets	0.60	0	0.65	0	0.575	0	0.65	0
Two static targets	1.313	0	1.548	0	1.30	0	1.55	0
Single moving target	N.A	2.34	N.A	N.A	0.85	2.43	N.A	N.A
Single moving target	N.A	5	N.A	N.A	0.85	4.87	N.A	N.A
One static target, one moving target	0.6	0	N.A	2.34	0.6	0	0.875	2.43

5.3. Speed Measurement

Figure 10 shows the measurement results for two different speeds. The speed of a single target (0.03 × 0.03 m²) is set to 2.34 m/s in Figure 10a, while Figure 10b shows the result for a single target (0.015 × 0.02 m²) moving at 5 m/s. The assembler, which contains the target mounted on a moving belt, is 0.26 m long and is placed 0.67 m from the receiving antenna. We can notice that both speed and distance can be identified, but due to the smaller size of the moving target, the received signal strength in Figure 10b is weaker than in Figure 10a.

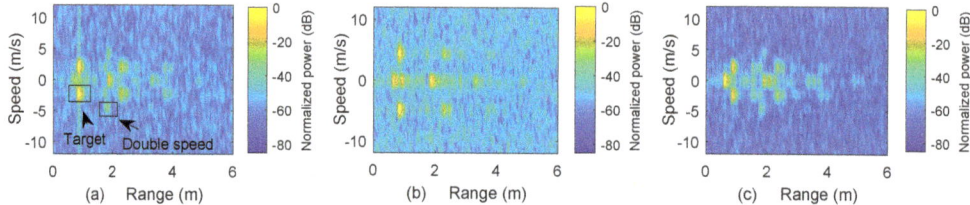

Figure 10. Range–speed plots for the speed measurements. (**a**) Range–speed plot of a single target moving with 2.34 m/s, whereas the moving assembler acts as a static unit. (**b**) Range–speed plot of the target in (**a**) with speed increased to 5 m/s. (**c**) Range–speed plot of a static target, placed at 0.6 m, and a moving target with speed of 2.34 m/s.

The result in Figure 10c are obtained by simultaneously placing a static target at 0.6 m and a moving target (0.015 × 0.02 m²) at 2.34 m/s located between 0.67 m to 0.9 m. It is observed that the setup can accurately measure the speed in different arrangements within the speed resolution of 0.79 m/s and the range of the static target within the range resolution of 0.042 m. Measurements for the speed values in our experiment required an interval of 1.96 ms to capture sinusoidal variations caused by the Doppler frequency. Multiple reflections from the moving target cause echoes with higher speeds, as indicated in Figure 10a. The ± speed in the results is due to the use of a double-sideband W-band mixer for up/down conversion in the experiment. The ± speed ambiguity can be removed by using single sideband devices such as IQ modulators.

5.4. Data Communication

To demonstrate the convergence of communication and sensing by using the same 16-QAM OFDM waveforms, we also measured data communication performance in terms of BER. A subframe wise processing is used, where 12 OFDM symbols form a subframe, and each OFDM symbol has 300 active subcarriers. In one subframe, four OFDM symbols are multiplexed with pilot subcarriers (4 × 50 pilot subcarrier in a subframe), and a code rate of 0.76 is used. In the experiment, we placed the receiver and transmitter in the LOS link distance of 0.6 m, with a bit rate of 8.08 Gbps. Due to non-ideal environment for the measurement such as limited dimensions of the lab and surrounding objects, multiple reflections of the transmitted signal arrive at the receiving antenna. Changes in the spectrum of the received signal in Figure 11 (red color) confirm the presence of the multiple reflections. Therefore, channel equalization is necessary to recover the transmitted data, which increases BER compared to AWGN channel [38]. We performed frequency domain equalization (zero-forcing) by using the transmitted pilot subcarriers for the channel estimation. As a result, a BER of 0.01 is recorded by comparing the transmitted and received data bits, when the average received SNR is around 15 dB, which is below the soft-decision forward error correction (SD-FEC) threshold of 1.5×10^{-2} [39]. As a comparison, a BER of 3.6×10^{-4} is obtained in the simulation due to the absence of the background reflections. A comparison of the received spectrum and the corresponding constellations is provided in Figure 11, for illustration purposes.

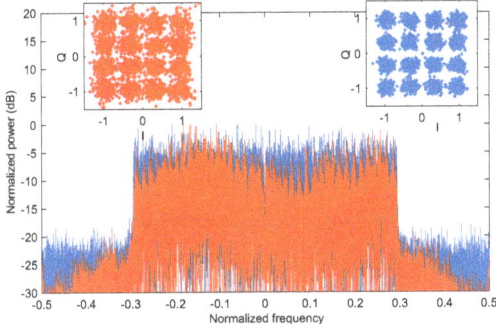

Figure 11. Data transmission results. The baseband spectrum and 16–QAM constellation, and red color for a LOS link of 0.6 m with background reflections. Blue color is the corresponding simulation result.

5.5. Comparison with Existing Works

A performance comparison of the proposed system is compared with the existing OFDM-based works in Table 4. In most of the experimental demonstrations for the convergence of communications and sensing, the OFDM waveform is evaluated for the sensing performance only. Since the speed resolution is linked with the observation time, higher speed resolution can be realized with longer observation intervals, as shown in Table 4.

Table 4. Performance comparison of the proposed converged system with the existing works.

Reference	Carrier Frequency	Bandwidth	Range Resolution	Observation Time	Speed Resolution	Data Rate
[34]	24 GHz	93 MHz	1.6 m	3.168 ms	1.97 m/s	20 Mbit/s
[31]	2.4 GHz	98.28 MHz	1.5 m	20 ms	4.2 m/s	Not evaluated
[32]	28 GHz	400 MHz	0.4 m	0.25 ms	N.A	Not evaluated
[28]	77 GHz	200 MHz	0.75 m	8 ms	0.48 m/s	Not evaluated
[29]	77 GHz	1.024 GHz	0.14 m	4.4 ms	0.38 m/s	Not evaluated
[30]	5.2 GHz	80 MHz	1.87 m	128 ms	0.22 m/s	Not evaluated
Proposed	97 GHz	3.9 GHz	0.04 m	1.9 ms	0.97 m/s	8.08 Gbit/s

6. Conclusions

We demonstrated a W-band simultaneous communication and sensing system operating at 97 GHz using a common 16-QAM OFDM waveform. The zero-delay-shift-based approach is proposed to overcome the sensing range limitation in the conventional OFDM-based sensing systems and enable range extension for the OFDM-based converged system. Both simulation and experimental measurements are performed in the W-band with a bandwidth of 3.9 GHz. Due to the large bandwidth available in the W-band, we achieve a sensing resolution of 0.042 m in range and 0.79 m/s in speed in the experiment. The target range well beyond the CPI is detected by using our proposed method. Furthermore, we also measure the 16-QAM OFDM communication performance, and the BER below the SD-FEC is achieved. The successful demonstration of the convergence of communication and sensing using the same waveform is a significant step towards future wireless applications.

Author Contributions: Conceptualization, N.M.I. and X.Y.; software, N.M.I., S.W. and L.Z.; data curation, N.M.I.; validation, N.M.I., Z.L., M.S. and H.Z.; investigation, N.M.I.; resources, X.Y.; writing—original draft preparation, N.M.I.; writing—review and editing, N.M.I. and X.Y.; visualization, N.M.I.; supervision, X.Y.; project administration, X.Y.; funding acquisition, X.Y. All authors have read and agreed to the published version of the manuscript.

Funding: This work was funded by the National Key Research and Development Program of China under Grant 2018YFB1801500, the Natural National Science Foundation of China under Grant 62101483, the Natural Science Foundation of Zhejiang Province under Grant LQ21F010015, and the Zhejiang Lab under Grant 2020LC0AD01.

Conflicts of Interest: The authors declare no conflict of interest.

References

1. Xiao, M.; Mumtaz, S.; Huang, Y.; Dai, L.; Li, Y.; Matthaiou, M.; Karagiannidis, G.K.; Björnson, E.; Yang, K.; Chih-Lin, I.; et al. Millimeter Wave Communications for Future Mobile Networks. *IEEE J. Sel. Areas Commun.* **2017**, *35*, 1909–1935. [CrossRef]
2. Mitola, J.; Guerci, J.; Reed, J.; Yao, Y.D.; Chen, Y.; Clancy, T.C.; Dwyer, J.; Li, H.; Man, H.; McGwier, R.; et al. Accelerating 5G QoE via Public-Private Spectrum Sharing. *IEEE Commun. Mag.* **2014**, *52*, 77–85. [CrossRef]
3. Gerla, M. Cognitive radios and the vehicular clouds. In Proceedings of the 1st ACM Workshop on Cognitive Radio Architectures for Broadband (CRAB), Miami, FL, USA, 4 October 2013; pp. 1–2.
4. Song, H.J.; Nagatsuma, T. *Terahertz Technologies Devices and Applications*; Taylor & Francis: Abingdon, UK; CRC Press: Boca Raton, FL, USA, 2015.
5. Paul, B.; Chiriyath, A.R.; Bliss, D. Survey of RF Communications and Sensing Convergence Research. *IEEE Access* **2017**, *5*, 252–270. [CrossRef]
6. Chiriyath, A.R.; Paul, B.; Bliss, D. Radar-Communications Convergence: Coexistence, Cooperation, and Co-Design. *IEEE Trans. Cogn. Commun. Netw.* **2017**, *3*, 1–12. [CrossRef]
7. Tavik, G.C.; Hilterbrick, C.L.; Evins, J.B.; Alter, J.J.; Crnkovich, J.G.; de Graaf, J.W.; Habicht, W.; Hrin, G.P.; Lessin, S.A.; Wu, D.C.; et al. The Advanced Multifunction RF Concept. *IEEE Trans. Microw. Theory Technol.* **2005**, *53*, 1009–1020. [CrossRef]
8. Jia, S.; Wang, S.; Liu, K.; Pang, X.; Zhang, H.; Jin, X.; Zheng, S.; Chi, H.; Zhang, X.; Yu, X. A Unified System with Integrated Generation of High-Speed Communication and High-Resolution Sensing Signals Based on THz Photonics. *J. Light. Technol.* **2018**, *36*, 4549–4556. [CrossRef]
9. Han, Y.; Ekici, E.; Kremo, H. Spectrum Sharing Methods for the Coexistence of Multiple RF Systems: A survey. *Ad Hoc Netw.* **2016**, *53*, 53–78. [CrossRef]
10. Saddik, G.N.; Singh, R.S.; Brown, E.R. Ultra-Wideband Multifunctional Communications/Radar System. *IEEE Trans. Microw. Theory Technol.* **2007**, *55*, 1431–1437. [CrossRef]
11. David, K.; Berndt, H. 6G Vision and Requirements: Is There Any Need for Beyond 5G? *IEEE Veh. Technol. Mag.* **2018**, *13*, 72–80. [CrossRef]
12. Letaief, K.B.; Chen, W.; Shi, Y.; Zhang, J.; Zhang, Y.A. The Roadmap to 6G: AI Empowered Wireless Networks. *IEEE Commun. Mag.* **2019**, *57*, 84–90. [CrossRef]
13. Hwang, T.; Yang, C.; Wu, G.; Li, S.; Li, G.Y. OFDM and Its Wireless Applications: A Survey. *IEEE Trans. Veh. Technol.* **2009**, *58*, 1673–1694. [CrossRef]
14. Salah, A.A.; Raja Abdullah, R.S.A.; Ismail, A.; Hashim, F.; Abdul Aziz, N.H. Experimental Study of LTE Signals as Illuminators of Opportunity for Passive Bistatic Radar Applications. *Electron. Lett.* **2014**, *50*, 545–547. [CrossRef]
15. Palmer, J.E.; Harms, H.A.; Searle, S.J.; Davis, L. DVB-T Passive Radar Signal Processing. *IEEE Trans. Signal Process.* **2013**, *61*, 2116–2126. [CrossRef]

16. Gogineni, S.; Rangaswamy, M.; Nehorai, A. Multi-modal OFDM waveform design. In Proceedings of the 2013 IEEE Radar Conference (RadarCon13), Ottawa, ON, Canada, 29 April–3 May 2013; pp. 1–5.
17. Li, Y.; Wang, X.; Ding, Z. Multi-Target Position and Velocity Estimation Using OFDM Communication Signals. *IEEE Trans. Commun.* **2020**, *68*, 1160–1174. [CrossRef]
18. Turlapaty, A.; Jin, Y.; Xu, Y. Range and velocity estimation of radar targets by weighted OFDM modulation. In Proceedings of the 2014 IEEE Radar Conference, Cincinnati, OH, USA, 19–23 May 2014; pp. 1358–1362.
19. Dokhanchi, S.H.; Shankar, M.B.; Nijsure, Y.A.; Stifter, T.; Sedighi, S.; Ottersten, B. Joint automotive radar-communications waveform design. In Proceedings of the 2017 IEEE 28th Symp. Personal Indoor and Mobile Radio Communication (PIMRC), Montreal, QC, Canada, 8–13 October 2017; pp. 1–7.
20. de Oliveira, L.G.; Nuss, B.; Alabd, M.B.; Diewald, A.; Pauli, M.; Zwick, T. Joint Radar-Communication Systems: Modulation Schemes and System Design. *IEEE Trans. Microw. Theory Technol.* **2021**. [CrossRef]
21. Barhumi, I.; Leus, G.; Moonen, M. Equalization for OFDM Over Doubly Selective Channels. *IEEE Trans. Signal Process.* **2006**, *54*, 1445–1458. [CrossRef]
22. Roque, D.; Siclet, C. Performances of Weighted Cyclic Prefix OFDM with Low-Complexity Equalization. *IEEE Commun. Lett.* **2013**, *17*, 439–442. [CrossRef]
23. Chiwoo, L.; Youngbin, C.; Jaeweon, C.; Panyuh, J.; Hyeonwoo, L. Novel OFDM transmission scheme to overcome caused by multipath delay longer than cyclic prefix. In Proceedings of the 2005 IEEE 61st Vehicular Technology Conference, Stockholm, Sweden, 30 May–1 June 2005; Volume 3, pp. 1763–1767.
24. Berger, C.R.; Demissie, B.; Heckenbach, J.; Willett, P.; Zhou, S. Signal Processing for Passive Radar Using OFDM Waveforms. *IEEE J. Sel. Top. Signal Process.* **2010**, *4*, 226–238. [CrossRef]
25. Searle, S.; Palmer, J.; Davis, L.; O'Hagan, D.W.; Ummenhofer, M. Evaluation of the ambiguity function for passive radar with OFDM transmissions. In Proceedings of the 2014 IEEE Radar Conference, Cincinnati, OH, USA, 19–23 May 2014; pp. 1040–1045.
26. Sturm, C.; Pancera, E.; Zwick, T.; Wiesbeck, W. A novel approach to OFDM radar processing. In Proceedings of the 2009 IEEE Radar Conference, Pasadena, CA, USA, 4–8 May 2009; pp. 1–4.
27. Mercier, S.; Bidon, S.; Roque, D.; Enderli, C. Comparison of Correlation-Based OFDM Radar Receivers. *IEEE Trans. Aerosp. Electron. Syst.* **2020**, *56*, 4796–4813. [CrossRef]
28. Pfeffer, C.; Feger, R.; Jahn, M.; Stelzer, A. A 77-GHz software defined OFDM radar. In Proceedings of the 2014 15th International Radar Symposium (IRS), Gdansk, Poland, 16–18 June 2014; pp. 1–5.
29. Schweizer, B.; Knill, C.; Schindler, D.; Waldschmidt, C. Stepped-Carrier OFDM-Radar Processing Scheme to Retrieve High-Resolution Range-Velocity Profile at Low Sampling Rate. *IEEE Trans. Microw. Theory Technol.* **2018**, *66*, 1610–1618. [CrossRef]
30. Schieler, S.; Schneider, C.; Andrich, C.; Döbereiner, M.; Luo, J.; Schwind, A.; Thomä, R.S.; Del Galdo, G. OFDM waveform for distributed radar sensing in automotive scenarios. In Proceedings of the 2019 16th European Radar Conference (EuRAD), Paris, France, 2–4 October 2019; pp. 225–228.
31. Barneto, C.B.; Riihonen, T.; Turunen, M.; Anttila, L.; Fleischer, M.; Stadius, K.; Ryynänen, J.; Valkama, M. Full-Duplex OFDM Radar with LTE and 5G NR Waveforms: Challenges, Solutions, and Measurements. *IEEE Trans. Microw. Theory Technol.* **2019**, *67*, 4042–4054. [CrossRef]
32. Barneto, C.B.; Rastorgueva-Foi, E.; Keskin, M.F.; Riihonen, T.; Turunen, M.; Talvitie, J.; Wymeersch, H.; Valkama, M. Millimeter-wave Mobile Sensing and Environment Mapping: Models, Algorithms and Validation. *IEEE Trans. Veh. Technol.* **2022**. [CrossRef]
33. Pham, T.M.; Bomfin, R.; Nimr, A.; Barreto, A.N.; Sen, P.; Fettweis, G. Joint Communications and Sensing Experiments Using mmWave Platforms. In Proceedings of the 2021 IEEE 22nd International Workshop on Signal Processing Advances in Wireless Communications (SPAWC), Lucca, Italy, 27–30 September 2021; pp. 501–505.
34. Sturm, C.; Wiesbeck, W. Waveform Design and Signal Processing Aspects for Fusion of Wireless Communications and Radar Sensing. *Proc. IEEE* **2011**, *99*, 1236–1259. [CrossRef]
35. Braun, M.; Sturm, C.; Niethammer, A.; Jondral, F.K. Parametrization of joint OFDM-based radar and communication systems for vehicular applications. In Proceedings of the 2009 IEEE 20th International Symposium on Personal, Indoor and Mobile Radio Communications, Tokyo, Japan, 13–16 September 2009; pp. 3020–3024.
36. Braun, K.M. OFDM Radar Algorithms in Mobile Communication Networks. Ph.D. Thesis, Karlsruhe Institute of Technology, Karlsruhe, Germany, 2014.
37. Braun, K.M.; Sturm, C.; Jondral, F.K. On the single-target accuracy of OFDM radar algorithms. In Proceedings of the 2011 IEEE 22nd International Symposium on Personal, Indoor and Mobile Radio Communications, Toronto, ON, Canada, 11–14 September 2011; pp. 794–798.
38. Cho, K.; Yoon, D. On the general BER expression of one- and two-dimensional amplitude modulations. *IEEE Trans. Commun.* **2002**, *50*, 1074–1080.
39. Otsuka, H.; Tian, R.; Senda, K. Transmission Performance of an OFDM-Based Higher-Order Modulation Scheme in Multipath Fading Channels. *J. Sens. Actuator Netw.* **2019**, *8*, 19. [CrossRef]

Article

Secure Transmission of Terahertz Signals with Multiple Eavesdroppers

Yuqian He [1], Lu Zhang [1], Shanyun Liu [2], Hongqi Zhang [1] and Xianbin Yu [1,2,*]

1. College of Information Science and Electronic Engineering, Zhejiang University, Hangzhou 310027, China
2. Zhejiang Lab, Hangzhou 310000, China
* Correspondence: xyu@zju.edu.cn

Abstract: The terahertz (THz) band is expected to become a key technology to meet the ever-increasing traffic demand for future 6G wireless communications, and a lot of efforts have been paid to develop its capacity. However, few studies have been concerned with the transmission security of such ultra-high-speed THz wireless links. In this paper, we comprehensively investigate the physical layer security (PLS) of a THz communication system in the presence of multiple eavesdroppers and beam scattering. The method of moments (MoM) was adopted so that the eavesdroppers' channel influenced by the PEC can be characterized. To establish a secure link, the traditional beamforming and artificial noise (AN) beamforming were considered as transmission schemes for comparison. For both schemes, we analyzed their secrecy transmission probability (STP) and ergodic secrecy capacity (ESC) in non-colluding and colluding cases, respectively. Numerical results show that eavesdroppers can indeed degrade the secrecy performance by changing the size or the location of the PEC, while the AN beamforming technique can be an effective candidate to counterbalance this adverse effect.

Keywords: THz communications; physical layer security; multiple eavesdroppers; beam scattering; artificial noise

1. Introduction

Wireless traffic volume has exponentially grown in recent years and wireless data rates exceeding 100 Gbit/s will be required in the coming decades [1]. As a result, new frequency spectra are demanded to fulfill the broad bandwidth requirements for future communication. Among others, the THz band (0.06–10 THz) is regarded as a promising candidate to enable ultra-fast and ultra-broadband data transmission [2–5]. Recently, THz wireless communication systems are under rapid development and many wireless transmissions exceeding 100 Gbit/s have already been demonstrated in laboratories and in field environments [6–11], which bring THz communication closer to reality. However, ultra-high-speed THz communications also pose major challenges to information security [12,13]. Once a malicious eavesdropper tries to intercept the signals, a vast amount of information will be leaked in the blink of an eye which is absolutely unacceptable, particularly in some sensitive fields such as the military and financial industry.

Security mechanisms exist at every layer of a network. Compared to the conventional upper-layer methods [14,15], physical-layer security (PLS) approaches [16–20] do not rely on the assumption that eavesdroppers have limited computational abilities and avoid distributing and managing secret keys [21–24]. In contrast to the broadcast nature of the microwave communication, highly directive THz waves are more prone to the blockage problem caused by the malicious eavesdropper [25,26]. Recently, researchers have comprehensively investigated the blocking effects of an illegal recipient and proposed a hybrid beamforming and reflecting scheme to eliminate the adverse effects [27,28]. In this environment, any eavesdropper intending to hide itself should control its size, otherwise, it may cast a detectable shadow and raise an alarm. Therefore, the performance of eavesdropping is restricted by the size of the illegal receiver. Alternatively, recent works have pointed out that an eavesdropper may put a tiny passive object instead of itself, like a metal cup or a mobile phone, inside the narrow beam to scatter THz electromagnetic waves [29,30]. By this mean, the bulky illegal receiver placed outside the THz beam can capture the information signal without raising an alarm, as a consequence. We note that the feasibility of this scheme has already been demonstrated in experiments in which the eavesdropper can even intercept a signal strength as good as that of the intended receiver. Nevertheless, all the aforementioned work using scatter (tiny passive object) only consider a single-eavesdropper scenario while a case with multiple eavesdroppers has not been investigated. The reflector in the narrow beam scattering THz waves to multiple eavesdroppers may bring a greater security threat to the THz communication system.

Compared to the single eavesdropper, multiple eavesdroppers can increase the occurrence of stronger attackers that are closer to the legitimate transmitter due to the random spatial distribution [31,32]. Additionally, multiple eavesdroppers may also combine their own observations and jointly process their received message, which will considerably degrade the secrecy performance [33–35]. From a practical point of view, multi-eavesdropper scenes will be widespread phenomenon in our future, since potential eavesdroppers in the ubiquitous Internet of Things (IoT) may be some curious legitimate devices belonging to different subsystems [36]. However, secrecy performance and secure transmission schemes in highly directive THz communication systems have not been yet analyzed in the presence of multiple eavesdroppers. Moreover, how to safeguard this point-to-point THz system against randomly located eavesdroppers is still unknown.

In this paper, we comprehensively investigated the secrecy performance of a highly directive THz communication link with multiple eavesdroppers. We established the received signal models with two different multi-antenna techniques, namely traditional beamforming and AN beamforming, as transmission schemes for comparison. We note that the received signal mode is affected by the fading channel, where both the large-scale and small-scale effects matter. We emulate the effect of perfect electric conductor (PEC) parameters on the received signal-to-noise ratio (SNR) of Eve in a multiple-eavesdropper environment. We derive the mathematical framework of the STP and ESC in both non-colluding and colluding cases, so that the secrecy performance of the THz wireless link can be characterized. The results show that Eves can successfully intercept a huge amount of information by changing some parameters, such as the density, size, and distance. As a countermeasure, Alice could consider the deployment of the AN beamforming technique to counterbalance the adverse effect of multiple eavesdroppers.

The rest of the paper is organized as follows. In Section 2, we introduce the system model in the presence of multiple eavesdroppers. In Section 3, we analyze the STP and ESC in non-colluding and colluding cases, respectively. In Section 4, we conduct simulation experiments and demonstrate the factors affecting the secrecy performance. In Section 5, we discuss how one may find the attackers. Finally, we give a brief conclusion in Section 6. Additionally, the important notations in this paper are listed in Table 1 to make this paper clearer.

Table 1. Parameter settings.

Side	Symbol	Parameter Setting	Value
Alice	P	Transmitting power	-10 dBm
	G_t	Antenna gain	25/27 dBi
	N	Antenna number	Independent variable
	η	Power allocation ratio	Independent variable

Table 1. Cont.

Side	Symbol	Parameter Setting	Value
Channel	R_S	Covering radius	10/15 m
	λ_p	Density of eavesdroppers	Independent variable
	N_E	Number of eavesdroppers	Independent variable
	a	Radius of cylinder	Independent variable
	d_2	Distance between Eve and PEC	Independent variable
	d_3	Distance between Alice and PEC	Independent variable
	m	Nakagami fading parameters	2
Bob	d_1	Distance between Alice and Bob	Independent variable
	G_r	Antenna gain	25/27 dBi
Other	c	Speed of light	3×10^8 m/s
	f	Frequency	Independent variable
	P_N	Noise power	−75 dBm
	W	Bandwidth	50 GHz
	N-C	Non-colluding case	-
	C	Colluding case	-

2. System Model

In this section, we first propose a security model for the THz system, in which two transmission schemes, namely traditional beamforming and AN beamforming, are adopted to prevent being overheard by multiple eavesdroppers. Then, the details of the highly directive channel of Bob h_B and the scatter channel of Eve h_E are investigated, respectively.

2.1. Signal Model

As shown in Figure 1a, a transmitter (Alice) sends a highly directive THz wave to the receiver (Bob) in the presence of multiple eavesdroppers (Eves). A PEC on the origin O is put inside this narrow beam between Alice and Bob. When there is an incident beam, PEC will scatter the THz signal to Eves in all directions (see Appendix A). We note that the PEC is located at the very edge of the THz beam with only a sliver of THz wave so it will not cast a detectable shadow in the receiver Bob. Additionally, the PEC is a cylinder which has the advantage of being able to scatter light in all directions, giving an attacker more flexibility. We model the locations of multiple eavesdroppers by the homogeneous Poisson point process (PPP) Φ in a circle region of radius R_S with a density λ_p, as shown in Figure 1b. The total number of Eves N_E in PPP is a random variable but the average number can be determined by $\overline{N_E} = \pi R_S^2 \lambda_p$. Due to the short transmission distance ($R_S < 15$ m) in an indoor environment, all receivers are supposed to be in a high SNR regime. Alice has N antennas while Bob and all the Eves use only one antenna each for reception.

When traditional beamforming is adopted, the received symbols at Bob and i-th Eve are, respectively, given by:

$$y_B = h_B x + n_B, \tag{1}$$

$$y_{E_i} = h_{E_i} x + n_E, \quad i = 1, 2, \cdots, N_E, \tag{2}$$

where h_B and h_{E_i} are both $1 \times N$ vectors denoting the channel between Alice and Bob and between Alice and the i-th Eve, respectively; N_E is the total number of eavesdroppers; $x = pu_x$ is the transmitted signal containing the beamforming vector p and signal u_x with useful information; n_B and n_E are i.i.d. additive white Gaussian noise with $n \sim \mathcal{CN}(0, \sigma_n^2)$. We assume that both Alice and Bob only know the CSI of h_B, while Eve knows both h_B and h_{E_i} perfectly, which is a more rigorous scenario for the security issue [37–39].

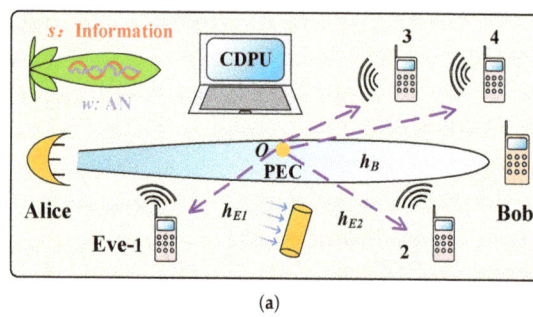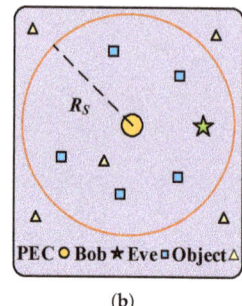

Figure 1. System model. (**a**) Alice transmits a highly directive THz signal x to Bob with or without AN w. A PEC (orange cylinder) located at the edge of beam can scatter the incident THz wave to Eves in all directions. (**b**) The spatial distribution of Eves is modeled as PPP in a circle region. The objects in this indoor scene can scatter THz signals.

With the introduction of AN beamforming, the transmitted THz signal x can be carefully designed as: $x = s + w$. The information signal $s = pu_s$, where the $N \times 1$ beamforming vector $p = h_B^\dagger/\|h_B\|$ and signal u_s with a variance of $\sigma_{u_s}^2$. The AN $w = Zv$, where the $N \times (N-1)$ matrix Z is the null space of vector h_B so that $h_B Z = 0$ while $h_{E_i} Z \neq 0$ and noise vector v contains $(N-1)$ random noise elements with a variance of σ_v^2. Consequently, the received signals of Bob and i-th Eve are, respectively, given by:

$$y_B = h_B(s + w) + n_B = h_B p u_s + n_B, \quad (3)$$

$$y_{E_i} = h_{E_i}(s + w) + n_E = h_{E_i} p u_s + h_{E_i} Z v + n_E. \quad (4)$$

The AN w passes through the channel h_{E_i} and finally develops into the additional noise $h_{E_i} w$. We stress that, despite the AN, the w on Alice's side is sent to both the i-th Eve and Bob, whereas on the receiving side, the AN only deteriorates the i-th Eve without impacting Bob. As we can see, there is additional noise $h_{E_i} Z v$ on Equation (4) while there is no extra term on Equation (3).

The total transmitter power $P = E[x^\dagger x] = \sigma_{u_s}^2 + (N-1)\sigma_v^2$, where $(\cdot)^\dagger$ denotes the conjugate transpose. We define η as the fraction of $\sigma_{u_s}^2$ to the total transmit power P. When $\eta = 1$, the AN beamforming is equivalent to traditional beamforming as the information signal is transmitted with the full power P. We note that η is an important design parameter that can optimize the secrecy performance.

2.2. Highly Directive Channel

The channel model of Bob h_B can be obtained as:

$$h_B = l_B s_B, \quad (5)$$

where l_B is the large-scale factor denoting the fixed pass loss and s_B is the small-scale random vector containing N elements. The l_B influenced by the free space pass loss (FSPL) and highly directive antennas is given by:

$$l_B = \frac{\lambda \sqrt{G_t G_r}}{4\pi d_1}, \quad (6)$$

where G_t and G_r, respectively, represent the antenna gains of Alice and Bob, and λ stands for the wavelength, and d_1 is the distance between Alice and Bob.

Unlike the conventional channel on the microwave band where the small-scale fading follows normal distribution, s_B on the THz band is usually represented by Nakagami-m distribution with the i-th element $s_{B_i} \sim Nakagami(m,1)$, which has recently been proven by experiments [40,41]. Finally, according to Equation (3), the signal-to-noise ratio (SNR) of Bob is given by:

$$\text{SNR}_B = S_B \frac{L_B P \eta}{\sigma_n^2}, \tag{7}$$

where $S_B \sim \text{Gamma}(mN, m)$ and $L_B = l_B^2$ are given by Equation (6).

2.3. Scatter Channel

The scatter channel of Eve h_E is given by:

$$h_{E_i} = l_{E_i} s_{E_i}, \tag{8}$$

where l_{E_i} and s_{E_i} are, respectively, the large-scale factor and small-scale random vector of i-th Eve. The l_E and s_E are totally different from l_B and s_B owing to the PEC between Alice and Bob. The PEC between Alice and Bob is a kind of material with infinite conductivity and zero electric field inside. When the incident field E_i strikes the surface of PEC, it provokes a surface current J_Z that generates a scattered field E_s and total reflection occurs. By adopting the method of moments (MoM) [42], the scatter field E_s around the PEC at i-th Eve is given by (see Appendix A):

$$E_s = \frac{-k\eta_0}{4\pi}\sqrt{\frac{\eta_0 P G_t}{kd_{2_i}}} \exp\{-j(kd_{2_i} - \frac{\pi}{4})\} \mathbf{C}^T \mathbf{A}^{-1} \mathbf{D}, \tag{9}$$

where k is the wave number, $\eta_0 \simeq 377\,\Omega$ is the intrinsic impedance of free space, d_{2_i} is the distance between the PEC and i-th Eve and the matrices $\mathbf{C}, \mathbf{A}, \mathbf{D}$ are determined by the shape, size, and location of the PEC. Here, we assume that the PEC is a cylinder with sufficient height. As such, we can denote the scattering coefficient $K(a, d_3) = \mathbf{C}^T \mathbf{A}^{-1} \mathbf{D}$, where a is the radius of PEC and d_3 is the distance between Alice and PEC. Therefore, the l_{E_i} can be derived as:

$$l_{E_i} = \sqrt{\frac{|E_s|^2}{2\eta_0} \frac{G_r \lambda^2}{4\pi P}} = \frac{\eta_0 \lambda K(a, d_3)}{8\pi} \sqrt{\frac{kG_t G_r}{2\pi d_{2_i}}}, \tag{10}$$

where we assume that Bob and all Eves have the same antenna gain G_r.

The scattering coefficient K is influenced by a and d_3. As shown in Figure 2, the THz wave nearly scatters uniformly around the PEC center ($d_2 \gg \lambda$, [42]) and the scattered field gradually fades along as it becomes farther away from the center. The scattering coefficient K increases with radius a and decreases with d_3, as we can see since the color in Figure 2b is deeper than that in Figure 2a.

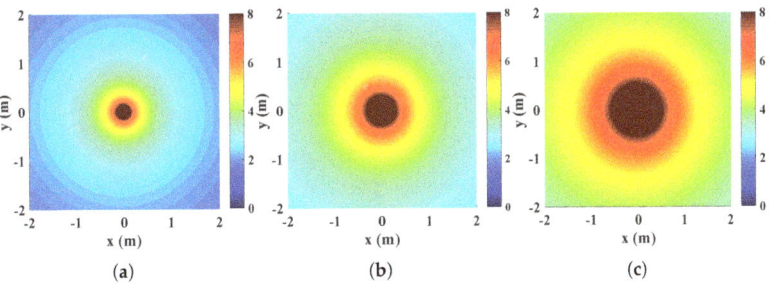

Figure 2. The scattered fields of PEC for (a) a = 20 mm, d_3 = 2 m (b) a = 40 mm, d_3 = 2 m (c) a = 40 mm, d_3 = 1.5 m. The maximum values were cut off at 8 since only a few values exceed it.

Unlike the main channel wherein a direct line-of-sight (LOS) link exists between Alice and Bob, Eve indirectly receives the signal information from non-line-of-sight (NLOS) transmission. Many rays will scatter from PEC and finally converge on Eve's side as each point on the surface of PEC can generate an electromagnetic field. As such, a tiny move of PEC or Eve may tremendously change the received signal strength. Therefore, we assume $s_{E_i} \sim Nakagami(1,1)$, which is also a *Rayleigh* distribution. Based on Equation (4), the SNR of *i*-th Eve is given by:

$$\text{SNR}_{E_i} = \frac{S_{E_i} L_{E_i} P \eta}{\frac{A L_{E_i} P(1-\eta)}{N_A - 1} + \sigma_n^2} \overset{(a)}{\leq} \phi S_{equal_i}, \tag{11}$$

where $A \sim Gamma(N-1)$, $S_{E_i} \sim Exp(1)$, $L_{E_i} = l_{E_i}^2$, the PDF of random variable S_{equal_i} is given by $f_{S_{equal}}(x) = \frac{N-1}{(1+x)^N}$ and $\phi = \frac{\eta(N-1)}{1-\eta}$, (a) holds for considering the worst-case situation where the normalized noise σ_n are arbitrarily small. Note that this approach was also taken in [16,35,37].

3. Secrecy Performance

In this section, we introduce STP and ECS which are both secrecy performance metrics. Then, we analyze the secrecy performance with and without AN in both non-colluding and colluding cases.

3.1. Performance Metrics

In the non-colluding case, the eavesdropper individually overhears the communication between Alice and Bob without any centralized processing. Therefore, the SNR of multiple eavesdroppers is given by $\text{SNR}_E = \max(\text{SNR}_{E_i})$, where SNR_{E_i} is defined in Equation (11). Whereas, in the colluding case, N_E Eves are capable of sending the information to a central data processing unit (CDPU) and jointly process their received information as shown in Figure 1a. Thus, the SNR of multiple eavesdroppers is given by $\text{SNR}_E = \sum_{i=1}^{N_E} \text{SNR}_{E_i}$. We adopt the following metrics to evaluate the secrecy performance of the proposed system.

Secure transmission probability (STP): STP is defined as a complementary element of secrecy outage probability (SOP) [31]. A supremum of the secrecy transmission rate R is determined by the difference of the main channel capacity $C_B = log(1 + \text{SNR}_B)$ and the wiretap channel capacity $C_E = log(1 + \text{SNR}_E)$. If secrecy transmission rates R are less than this supremum $C_S = C_B - C_E$, a secure transmission can be realized, otherwise, a secrecy outage occurs. The STP in non-colluding and colluding cases are, respectively, defined as:

$$P(C_S > R) = \prod_{E_i \in \Phi} P(\frac{1+\text{SNR}_B}{1+\text{SNR}_{E_i}} > 2^R), \tag{12}$$

$$P(C_S > R) = P(\frac{1+\text{SNR}_B}{1+\sum \text{SNR}_{E_i}} > 2^R). \tag{13}$$

Ergodic secrecy capacity (ESC): ESC is defined as the average transmission rate of the confidential message, which is formulated as:

$$\overline{C}_S = E[C_S] = \int_0^\infty P(C_S > R) dR. \tag{14}$$

In practice, ECS is used to describe the fast fading channel while STP for a slow fading channel. However, the numerical value of ECS is still determined by the STP. As long as we obtain the STP, the ESC can be simply calculated by its integration.

3.2. Non-Colluding Eavesdroppers

In a non-colluding eavesdroppers scenario, we investigated the STP for traditional beamforming ($\eta = 1$) and AN beamforming ($\eta \neq 1$). When traditional beamforming is adopted, we derived the exact value of STP, whereas AN is introduced, and we calculated the lower bound of STP which is a rigorous assumption and common practice [16,37].

We denote the STP for traditional beamforming as P_1 and for AN beamforming as P_2, respectively. Based on Equation (12), P_1 is given by:

$$P_1 = \prod_{E_i \in \Phi} P(\frac{1+\text{SNR}_B}{1+\text{SNR}_{E_i}} > 2^R) \stackrel{(b)}{=} E_{S_B}\{\exp(-2\pi\lambda_p \int_0^{R_S} P(S_E > \frac{S_B L_B}{2^R L_E})\rho d\rho)\}, \quad (15)$$

where (b) holds for $\text{SNR}_B \gg 1$, $\text{SNR}_E \gg 1$ and the *probability generating functional lemma* (PGFL, ref. [43]) over PPP.

By denoting $u = kG_tG_r(\eta_0 K\lambda)^2/128\pi^3$, we have $L_E = u\frac{1}{d_2^2}$. As $S_E \sim E(1)$, the Equation (15) can finally be derived as:

$$P_1 = E_{S_B}\{\exp(2\pi\lambda_p((vR_S+v^2)e^{-\frac{R_S}{v}} - v^2))\}, \quad (16)$$

where $v = u2^R/S_B L_B$.

Similarly to the calculation of P_1 and by denoting $\beta = \frac{2^R \sigma_n^2}{PL_B}$, $P_2(C_S > R)$ is given by:

$$P_2 = \prod_{E_i \in \Phi} P(\frac{1+\text{SNR}_B}{1+\text{SNR}_{E_i}} > 2^R) \stackrel{(c)}{=} E_{S_B}\{\exp(\frac{-\pi R_S^2 \lambda_p}{(1+\frac{S_B\eta-\beta}{\beta\phi})^{N-1}})\}, \quad (17)$$

where (c) holds for $\text{SNR}_B \gg 1$ and the PGFL over PPP.

3.3. Colluding Eavesdroppers

In colluding case, we denote the STP without AN as P_3 and with AN as P_4, respectively. The STP $P_3(C_S > R)$ is given by:

$$P_3 = P(\frac{1+\text{SNR}_B}{1+\sum \text{SNR}_{E_i}} > 2^R) = P(S_B > \frac{2^R \sum_{E_i \in \Phi} S_{E_i} L_{E_i}}{L_B}). \quad (18)$$

We let $I_1 = \sum_{E_i \in \Phi} S_{E_i} L_{E_i}$ and thus P_3 can be modified as:

$$P_3 = \int_0^\infty P(S_B > p_1 i)f_{I_1}(i)di \stackrel{(d)}{=} \sum_{b=0}^{mN-1} m_b p_1^b (-1)^b \mathcal{L}^{(b)}\{f_{I_1}(i)\}(mp_1), \quad (19)$$

where $f_{I_1}(i)$ is the probability density function (PDF) of I_1 and $p_1 = 2^R/L_B$, (d) holds for $S_B \sim \text{Gamma}(mN, m)$ and the complementary cumulative distribution function (CCDF) of S_B is given by $F_{S_B}^c = e^{-mx}\sum_{b \in B} m_b x^b$, where $m_b = \frac{m^b}{b!}$ and $b \sim (0, mN-1)$. The Laplace transformation $\mathcal{L}\{f_{I_1}(i)\}(mp_1)$ of function $f_{I_1}(i)$ is given by:

$$\mathcal{L}\{f_I\}(p_1) = \exp\{-2\pi\lambda_p p_1 u R_S\}(1+\frac{R_S}{p_1 u})^{2\pi\lambda p_1^2 u^2}. \quad (20)$$

By adopting *Bruno's formula* [44], we can obtain the n-degree derivation of $\mathcal{L}\{f_{I_1}(i)\}(p_1)$ as:

$$\mathcal{L}^{(n)}\{f_{I_1}\}(p_1) = \sum \frac{n!}{b_1!\cdots b_n!}e^{f(p_1)}\prod_{j=1}^n (\frac{f^{(j)}(p_1)}{j!})^{b_j}, \quad (21)$$

where the sum is over all the solutions $b_1, \cdots, b_n \geq 0$ to $b_1 + 2b_2 + \cdots + nb_n = n$. By denoting $w = R_S/u$, $c_1 = 2\pi\lambda_p$, $c_2 = 1 + \frac{w}{p_1}$, $c_3 = \frac{1}{p_1+w} - \frac{1}{p_1}$, $c_4 = \frac{1}{p_1^2} - \frac{1}{(p_1+w)^2}$, $f(p_1)$ and $f^{(j)}(p_1)$ are given by:

$$f(p_1) = c_1(p_1^2 u^2 \ln c_2 - p_1 u R_S), \quad (22a)$$

$$f^{(1)}(p_1) = c_1(2p_1 u^2 lnc_2 + p_1^2 u^2 c_3 - uR_S), \tag{22b}$$

$$f^{(2)}(p_1) = c_1(2u^2 lnc_2 + 4p_1 u^2 c_3 + p_1^2 u^2 c_4), \tag{22c}$$

$$f^{(j>2)}(p_1) = c_1 u^2 \sum_{kk=0}^{2} C_2^{kk}(-1)^{j-kk} \frac{j!}{(j-kk)} p_1^{2-kk}(\frac{1}{p_1^{j-kk}} - \frac{1}{(p_1+w)^{j-kk}}). \tag{22d}$$

When AN beamforming is introduced, $P_4(C_S > R)$ is given by:

$$P_4 = \int_0^\infty P(S_B > p_2 i) f_{I_2}(i) di = P(S_B > \frac{\beta(1 + \sum_{E_i \in \Phi} \phi S_{equal_i})}{\eta}). \tag{23}$$

We let $I_2 = 1 + \sum_{E_i \in \Phi} \phi S_{equal_i}$ and thus P_4 can be rewritten as:

$$P_4 = \int_0^\infty P(S_B > p_2 i) f_{I_2}(i) di = \sum_{b=0}^{mN-1} m_b p_2^b (-1)^b \mathcal{L}^{(b)}\{f_{I_2}(i)\}(mp_2), \tag{24}$$

where $f_{I_2}(i)$ is the PDF of I_2 and $p_2 = \frac{\beta}{\eta}$. As long as we obtain $\mathcal{L}\{f_{I_2}(i)\}(mp_2)$, P_4 can be calculated. The Laplace transformation $\mathcal{L}\{f_{I_2}(i)\}(p_2)$ is given by:

$$\mathcal{L}\{f_{I_2}\}(p_2) = exp\{-p_2 - N_E + q_1 q_2\}, \tag{25}$$

where $q_1 = exp(p_2\phi) E_N(p_2\phi)$, $E_N(x) = \int_1^\infty \frac{e^{-xt}}{t^N} dt$ is the N-degree exponential integral and $q_2 = N_E(N-1)$. As such, the n-degree of $\mathcal{L}\{f_{I_2}(i)\}(p_2)$ is given by:

$$\mathcal{L}^{(n)}\{f_{I_2}\}(p_2) = \sum \frac{n!}{b_1! \cdots b_n!} e^{g(p_2)} \prod_{j=1}^{n} (\frac{g^{(j)}(p_2)}{j!})^{b_j}, \tag{26}$$

where $g(p_2)$ and $g^{(j)}(p_2)$ are given by:

$$g(p_2) = -p_2 - N_E + q_1 q_2, \tag{27a}$$

$$g^{(1)}(p_2) = -1 + q_2 \phi e^{k\phi}(E_N - E_{N-1}), \tag{27b}$$

$$g^{(j \geq 2)}(p_2) = q_2 \phi^j e^{k\phi} \sum_{jj=0}^{j} C_j^{jj}(-1)^{jj} E_{N-jj}. \tag{27c}$$

4. Security Analysis

In what follows, we describe Eve's strategies to degrade the secrecy performance with a PEC, and then we show the function of AN as a countermeasure to resist the multiple eavesdroppers. Meanwhile, power allocation as a significant parameter of AN beamforming is also analyzed. Table 1 shows the parameter settings.

4.1. Eve's Attack

The intensity of Eves' attack is affected by the spatial distribution. In Figure 3a, when we compare the blue line with the red and yellow line, respectively, we find that the covering radius R_S has little effect on STP while density λ_p significantly reduces the STP. However, in Figure 3b, both the value of R_S and λ_p have significant impacts on the STP. The reason is that the SNR of multiple eavesdroppers in the non-colluding case depends on the 'nearest' Eve which has the best channel quality while the SNR in the colluding case only depends on the total number. The parameter R_S can barely increase the chance of the 'nearest' Eve as the THz transmit power quickly attenuates with the distance but indeed increases the total number of them. Therefore, from Eves' perspective, they have to focus on 'a better channel' or 'a better location' rather than the total number in a non-colluding case, as we can see the STP of the case when $\overline{N_E} = 17$ performs even better than the STP when $\overline{N_E} = 7$ in Figure 3a.

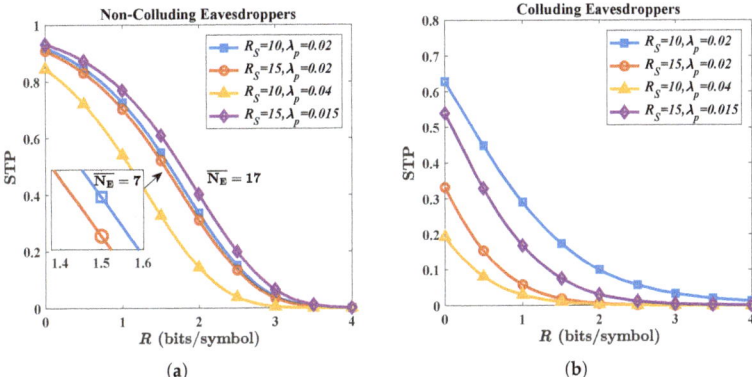

Figure 3. Secure transmission probability (STP) under different R_S and λ_p for (**a**) non-colluding case and (**b**) colluding case. Parameters are given by: G = 25 dBi; N = 5; f = 300 GHz; and P = −10 dBm.

In Figure 4, we use normalized secrecy capacity [30] to show the extent to which Eves reduce the secrecy capacity in non-colluding and colluding cases, respectively. It is shown that for d_2 = 20, the existence of Eves reduces the original capacity by 20% in non-colluding case and by nearly 40% in colluding case.

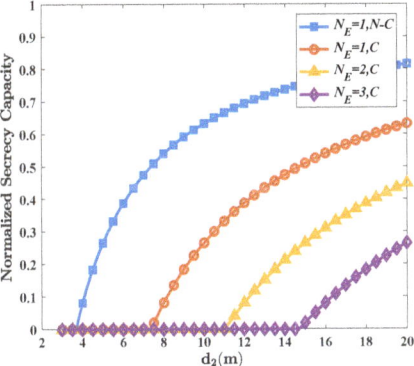

Figure 4. The normalized secrecy capacity as a function of d_2 in the non-colluding and colluding cases. Here, all the eavesdroppers have the same distance d_2 to the PEC and the channel fading is ignored. Other parameters are given by: G = 25 dBi; f = 300 GHz; P = −10 dBm; R_S = 15 m; and d_3 = 1 m.

Eve can move the PEC closer to Alice to strengthen the attack. In Figure 5, we find that the ESC monotonically increases with d_3 (PEC) while decreasing with the d_1 (Bob). In addition, the parameters d_1 and d_3 may have interacted with each other. For example, for d_1 = 3, a unit increase in d_3 will give birth to the improved ESC by ΔESC = 1.45. For d_1 = 5, ΔESC becomes 1.95. That is to say, $d_3(d_1)$ may exhibit a different effect when the other factor changes. Furthermore, if PEC is located in the midpoint between Alice and Bob, the ESC will not change significantly with the increase in d_1, as we can see that the white line in Figure 5 nearly remains unchanged at ESC = 4.85.

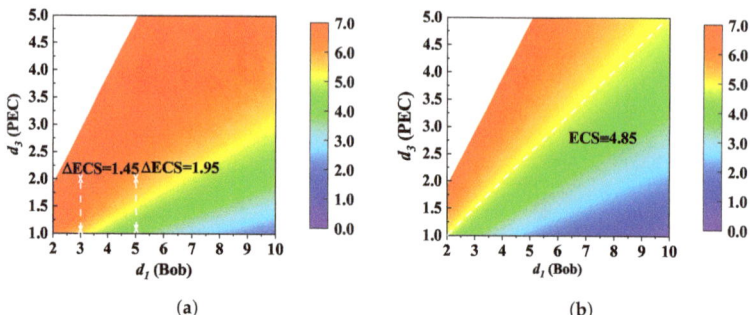

Figure 5. Ergodic secrecy capacity (ESC) as a function of d_1 (Alice–Bob) and d_3 (Alice–PEC) for (a) non-colluding eavesdroppers and (b) colluding eavesdroppers. Other parameters are given by: $G = 25$ dBi; $N = 3$; $f = 300$ GHz; $P = -10$ dBm; $R_S = 15$ m; and $\lambda_p = 0.015$.

Eve can increase the size of PEC to strengthen the attack. In Figure 6a, we find that the STP will decrease when the radius a rises from 20 mm to 40 mm, regardless of whether it is in the non-colluding case or in the colluding case. As shown in Figure 2, as a grows from 20 mm to 40 mm, the electromagnetic field around the PEC will be augmented and hence Eves obtains better signal quality. Additionally, we find that Eves benefit from increasing a to various degrees when the location of PEC d_3 changes. For $d_3 = 5$ m, as shown in Figure 6b, reducing a from 10 mm to 50 mm will lead to an ESC decrease of 36%. For $d_3 = 1$ m, however, reducing a from 10 to 50 decreases the ESC by 87% to nearly 0 which means that Eves can almost intercept all the information. Since being too near to Alice will increases the risk of being detected, Eve's strategy is to select a proper size and optimal location in such a way she can obtain as good a signal strength as possible and hide herself simultaneously.

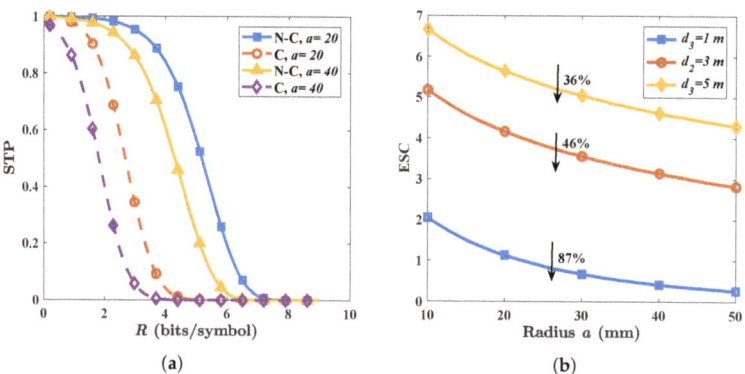

Figure 6. (a) Influence of radius a on STP. (b) ESC versus radius a under different PEC location d_3. The solid line describes the non-colluding case while the dashed line describes the colluding case. Other parameters are given by: $G = 25$ dBi; $N = 3$; $f = 300$ GHz; $P = -10$ dBm; $R_S = 15$ m; and $\lambda_p = 0.04$.

4.2. AN as a Countermeasure

We find that the AN beamforming can compensate for the detriment of multiple eavesdroppers. As shown in Figure 7, the increase in λ_p causes an STP ($P(C_S \geq 0)$) reduction from 0.85 to 0.75 and 0.5 to 0.1, respectively. However, with the introduction of AN in the non-colluding case, the STP ($P(C_S \geq 0)$) rises to 0.95, leading to an improvement of nearly 27%. In the colluding case, the STP rises to 0.7, corresponding to an improvement of 600%. It is noteworthy that the detriment of multiple eavesdroppers in the colluding case is more than that in the non-colluding case. In the non-colluding case, for $R > 2$, the STP with AN beamforming ($\lambda_p = 0.02$) is higher than that with traditional beamforming ($\lambda_p = 0.01$). However, in the colluding case, the situation is reversed for $R > 2$ which means that colluding eavesdroppers cause greater damage to transmission security.

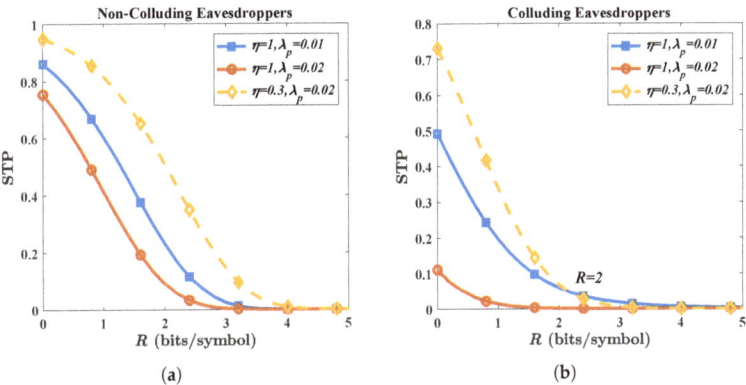

Figure 7. The benefit of AN on STP for (**a**) a non-colluding case; and a (**b**) colluding case. Other parameters are given by: $G = 25$ dBi, $N = 3$, $f = 300$ GHz, $P = -10$ dBm, $R_S = 15$ m, $\eta = 0.3$.

In Figure 8, we find that the optimal η depends on both density λ_p and the number of antennas N. For $\lambda_p = 0.01$ (blue and yellow line), the optimal η in the non-colluding and colluding cases is 0.28 and 0.22, respectively, which are larger than 0.21 and 0.15 for $\lambda_p = 0.02$. More Eves around PEC signify stronger information attacks. Therefore, Alice must allocate more transmission power to AN to resist the adverse effect of the added Eves. Additionally, the optimal value of η increases with N. As shown in the inset, the optimal η are 0.34 and 0.27 for $N = 6$ while 0.28 and 0.22 for $N = 2$. We stress that only Bob benefits from the increase in antennas since the transmitter maximizes the signal strength to Bob and the signal power at Eves' side remains unchanged.

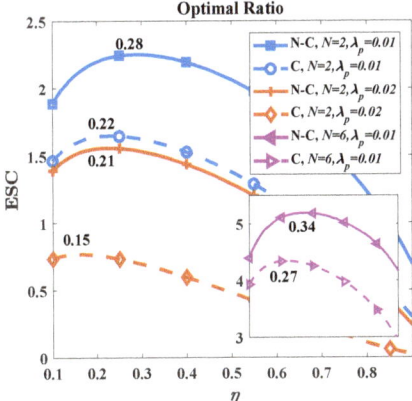

Figure 8. The optimal η under different λ_p and N. The solid line describes non-colluding cases while the dashed line describes colluding cases. The main figure for $N = 2$ while the inset for $N = 6$. Other parameters are given by: $G = 27$ dBi; $f = 300$ GHz; $P = -10$ dBm; and $R_S = 15$ m.

In Figure 9, we find that the ESC decreases with the f while increasing with the P when $\eta \neq 1$. For a system without AN, the ESC will not be influenced by P since SNR_B and SNR_E benefit from them to the same extent, as shown by Equations (7) and (11). However, with the introduction of AN, P can no longer influence the supremum of SNR_E but still impacts SNR_B. Additionally, we also find P and f cannot significantly change the optimal η. In Figure 9a,b, the optimal η varies in the ranges of 0.27~0.31 and 0.26~0.3 with standard deviations (STD) of 1.13×10^{-2} and 1.14×10^{-2}, respectively, lending to a tiny change. We note that despite Figure 9 only showing a non-colluding case, the same rule can also be applied to the colluding scenario.

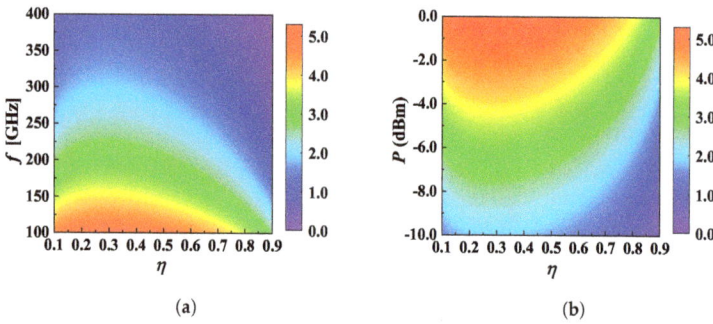

Figure 9. Secrecy performance in a non-colluding case. (**a**) The ECS as a function of η and f with $P = -10$ dBm; (**b**) The ECS as a function of η and P with f = 300 GHz. Other parameters are given by: G = 25 dBi, N = 3, R_S = 15 m, λ_p = 0.02.

5. Discussion

In practice, the first step to guarantee transmission security is to determine whether attackers exist instead of determining how to resist attackers. Therefore, before using unique techniques (such as AN), we should adopt a specific measure to detect the existence of an attacker, otherwise, many resources will be wasted. Recent work in [30] can successfully distinguish the suspicious objects from the ordinary environment through measuring the incoming signal. Here, we consider the possibility of increasing the beam directivity or enlarging the aperture of the receiver to guarantee the security. In this paper, the diameter of the THz beam is larger than the aperture of the receiver. Thus, Eves can utilize the edge of the beam to realize an attack. However, if the receiver has the ability to capture all of the transmitted THz wave without any leakage, any eavesdroppers trying to put an object in the beam will cause an extensive power reduction on Bob' side. In this case, if Eves still wants to implement an attack, she needs to either utilize the misalignment effect between Alice and Bob which may also induce a leakage or pretend to be irrelevant moving objects. Nevertheless, either way, Eves' strategy to implement an attack would be significantly more complicated and harder to implement. Another purpose of increasing the directivity is to resist the interference. Transceivers on the same unlicensed bandwidth may have interacted with each other. Additionally, jammers can also take advantage of this large bandwidth in the THz band for interference [45]. Increasing their directivity gains can make irrelevant transceivers and jammers either less effective or need to increase their transmit power.

In some cases, Eves are not afraid of being found because they are intended to block the signal power of Bob (reduce the secrecy capacity at the same time). As a countermeasure, multiple IRS-assisted THz systems with opportunistic connectivity may be a choice since Alice can choose different ways to transmit the signal and design unique beamforming schemes to maximize the secrecy rate performance. Researchers have found that opportunistic connectivity [46] with well-designed beamforming schemes can significantly boost the secrecy rate performance and reduce blocking probability.

6. Conclusions

In this paper, we investigated the secure transmission of THz waves in the indoor environment against randomly distributed eavesdroppers. We established the PLS model for this THz communication system, where Bob's channel is featured by a highly directive beam while Eve's channel scatters THz waves. Particularly, we characterize both channels with stochastic small-scale fading in order to accommodate the random variation in practice such as scattering on aerosols or the movement of objects. The security performance of traditional beamforming and AN beamforming in both non-colluding and colluding cases are analyzed by deriving the STP and ESC. Based on our analysis, we reveal that Eves can indeed take different strategies to degrade the secrecy performance, for instance, by changing the size or the distance of the scatter and increasing the density. To deal with this issue, an AN beamforming technique with a well-designed power allocation can be an effective candidate to counterbalance this adverse effect. Our study can not only serve as an inspiration for eavesdropping scenes but also for a widespread network scenario. Future work may extend this point-to-point communication scene to an indoor THz wireless local area networks (WLANs) which seem more appealing.

Author Contributions: Design, fabrication, and data analysis, Y.H. and X.Y.; software, Y.H. and L.Z.; writing—original draft preparation, Y.H.; writing—review and editing, S.L. and H.Z.; supervision, X.Y. All authors have read and agreed to the published version of the manuscript.

Funding: This work was supported by the National Key Research and Development Program of China (2020YFB1805700), in part by the Natural National Science Foundation of China under Grant 62101483, in part by the Natural Science Foundation of Zhejiang Province under Grant LQ21F010015, the Fundamental Research Funds for the Zhejiang Lab (no. 2020LC0AD01), the State Key Laboratory of Advanced Optical Communication Systems and Networks of Shanghai Jiao Tong University and in part by Zhejiang Lab (NO. 2020LC0AA03).

Data Availability Statement: Not applicable.

Conflicts of Interest: The authors declare no conflict of interest.

Appendix A

When the incident field E_i strikes the surface of PEC, it provokes surface current J_Z on PEC, which in turn generates a scattered field E_s which is given by:

$$E_s(\rho) = -\frac{k\eta_0}{4} \int_C J_Z(\rho') H_0^{(2)}(k|\rho - \rho'|) dS', \tag{A1}$$

where ρ is the field point on the plane, ρ' is the source point on the surface and $H_0^{(2)}$ is the Hankel function of the second kind of zero order. The integral in Equation (A1) is along the surface C which is divided into N_C segments. According to the property of PEC, the incident field of segment n $E_i(\rho'_n)$ is given by:

$$E_i(\rho'_n) = -\frac{k\eta_0}{4} \sum_{m=1}^{N} J_Z(\rho'_m) H_0^{(2)}(k|\rho'_n - \rho'_m|) \Delta C_m, \tag{A2}$$

where ρ'_n, ρ'_m is, respectively, the midpoint of segment n and m and ΔC_m is the length of segment m. By applying Equation (A2) to all the segments, there are totally N_C equations and all the equations can be cast in matrix form as:

$$\begin{bmatrix} E_i(\rho'_1) \\ \cdots \\ E_i(\rho'_{N_C}) \end{bmatrix} = \begin{bmatrix} A_{11} & \cdots & A_{1N_C} \\ \cdots & \cdots & \cdots \\ A_{N_C 1} & \cdots & A_{N_C N_C} \end{bmatrix} \begin{bmatrix} J(\rho'_1) \\ \cdots \\ J(\rho'_{N_C}) \end{bmatrix}, \tag{A3}$$

where the elements of impedance matrix **A** are influenced by the PEC itself and the incident field of segment n $E_i(\rho'_n)$ is also given by $E_i(\rho'_n) = \sqrt{2\eta_0 P G_t / 4\pi D_n^2}$, where $D_n = d_3 +$

$a\cos\theta_n$ is the distance between Alice and the segment n. Finally, we can calculate the scatter field E_s by substituting $E_i(\rho'_n)$ and Equation (A3) into Equation (A2):

$$E_s(\rho) = \frac{-k\eta_0}{4} \begin{bmatrix} H_0^{(2)}(k|\rho - \rho'_1|)\Delta C_1 \\ \cdots \\ H_0^{(2)}(k|\rho - \rho'_{N_C}|)\Delta C_{N_C} \end{bmatrix}^T \mathbf{A}^{-1} \begin{bmatrix} E_i(\rho'_1) \\ \cdots \\ E_i(\rho'_{N_C}) \end{bmatrix}$$

$$\stackrel{(e)}{=} \frac{-k\eta_0}{4\pi} \sqrt{\frac{\eta_0 P G_t}{k d_2}} \exp\{-j(k d_2 - \frac{\pi}{4})\} \mathbf{C}^T \mathbf{A}^{-1} \mathbf{D},$$

(A4)

where $\mathbf{C} = [\Delta C_1 \cdots \Delta C_{N_C}]$, $\mathbf{D} = [1/D_1 \cdots \Delta 1/D_{N_C}]^T$, (e) holds for $kd_2 \gg 1$ in the THz band so that approximations can be made with $|\rho - \rho'| \approx d_2$.

References

1. Nagatsuma, T.; Ducournau, G.; Renaud, C.C. Advances in terahertz communications accelerated by photonics. *Nat. Photonics* **2016**, *10*, 371–379. [CrossRef]
2. Chen, S.; Liang, Y.C.; Sun, S.; Kang, S.; Cheng, W.; Peng, M. Vision, requirements, and technology trend of 6G: How to tackle the challenges of system coverage, capacity, user data-rate and movement speed. *IEEE Wirel. Commun.* **2020**, *27*, 218–228. [CrossRef]
3. Huq, K.M.; Kazi, M.S.; Busari, S.A.; Rodriguez, J.; Frascolla, V.; Bazzi, W.; Sicker, D.C. Terahertz-enabled wireless system for beyond-5G ultra-fast networks: A brief survey. *IEEE Netw.* **2019**, *33*, 89–95. [CrossRef]
4. Zhang, Z.; Xiao, Y.; Ma, Z.; Xiao, M.; Ding, Z.; Lei, X.; Karagiannidis, G.K.; Fan, P. 6G wireless networks: Vision, requirements, architecture, and key technologies. *IEEE Veh. Technol. Mag.* **2019**, *14*, 28–41. [CrossRef]
5. Han, C.; Bicen, A.O.; Akyildiz, I.F. Multi-ray channel modeling and wideband characterization for wireless communications in the terahertz band. *IEEE Trans. Wirel. Commun.* **2015**, *14*, 2402–2412. [CrossRef]
6. Yu, Y.; Jia, S.; Hu, H.; Galil, M.; Morioka, T.; Jepsen, P.U.; Oxenløwe, L.K. 160 Gbit/s photonics wireless transmission in the 300–500 GHz band. *Apl Photonics* **2016**, *1*, 081301. [CrossRef]
7. Jia, S.; Pang, X.; Ozolins, O.; Yu, X.; Hu, H.; Yu, J.; Guan, P.; Ros, F.D.; Po, S.; Jacobsen, G.; et al. 0.4 THz photonic-wireless link with 106 Gb/s single channel bitrate. *J. Lightw. Technol.* **2018**, *36*, 610–616. [CrossRef]
8. Zhang, H.; Zhang, L.; Wang, S.; Lu, Z.; Yang, Z.; Liu, S.; Qiao, M.; He, Y.; Pang, X.; Zhang, X.; et al. Tbit/s multi-dimensional multiplexing THz-over-fiber for 6G wireless communication. *J. Lightw. Technol.* **2021**, *39*, 5783–5790. [CrossRef]
9. Wang, S.; Lu, Z.; Li, W.; Jia, S.; Zhang, L.; Qiao, M.; Pang, X.; Idrees, N.; Saqlain, M.; Gao, X.; et al. 26.8-m THz wireless transmission of probabilistic shaping 16-QAM-OFDM signals. *APL Photonics* **2020**, *5*, 056105. [CrossRef]
10. Harter, T.; Füllner, C.; Kemal, J.N.; Ummethala, S.; Steinmann, J.L.; Brosi, M.; Hesler, J.L.; Bründermann, E.; Müller, A.-S.; Freude, W.; et al. Generalized Kramers–Kronig receiver for coherent terahertz communications. *Nat. Photonics* **2020**, *14*, 601–606. [CrossRef]
11. Idrees, N.M.; Lu, Z.; Saqlain, M.; Zhang, H.; Wang, S.; Zhang, L.; Yu, X. A W-Band Communication and Sensing Convergence System Enabled by Single OFDM Waveform. *Micromachines* **2022**, *13*, 312. [CrossRef]
12. Yang, P.; Xiao, Y.; Xiao, M.; Li, S. 6G wireless communications: Vision and potential techniques. *IEEE Netw.* **2019**, *33*, 70–75. [CrossRef]
13. Tariq, F.; Khandaker, M.R.; Wong, K.K.; Imran, M.A.; Bennis, M.; Debbah, M. A speculative study on 6G. *IEEE Wirel. Commun.* **2020**, *27*, 118–125. [CrossRef]
14. Zou, Y.; Zhu, J.; Wang, X.; Hanzo, L. A survey on wireless security: Technical challenges, recent advances, and future trends. *Proc. IEEE* **2016**, *104*, 1727–1765. [CrossRef]
15. Yang, N.; Wang, L.; Geraci, G.; Elkashlan, M.; Yuan, J.; Renzo, M.D. Safeguarding 5G wireless communication networks using physical layer security. *Nat. Photonics* **2015**, *53*, 20–27. [CrossRef]
16. Goel, S.; Negi, R. Guaranteeing secrecy using artificial noise. *IEEE Trans. Wirel. Commun.* **2008**, *7*, 2180–2189. [CrossRef]
17. Li, B.; Zhang, M.; Rong, Y.; Han, Z. Artificial Noise-Aided Secure Relay Communication With Unknown Channel Knowledge of Eavesdropper. *IEEE Trans. Wirel. Commun.* **2021**, *20*, 3168–3179. [CrossRef]
18. Yang, F.; Zhang, K.; Zhai, Y.; Quan, J.; Dong, Y. Artificial Noise Design in Time Domain for Indoor SISO DCO-OFDM VLC Wiretap Systems. *J. Lightw. Technol.* **2021**, *39*, 6450–6458. [CrossRef]
19. Xu, W.; Li, B.; Tao, L.; Xiang, W. Artificial Noise Assisted Secure Transmission for Uplink of Massive MIMO Systems. *IEEE Trans. Veh. Technol.* **2021**, *70*, 6750–6762. [CrossRef]
20. Ding, X.; Song, T.; Zou, Y.; Chen, X.; Hanzo, L. Security-reliability tradeoff analysis of artificial noise aided two-way opportunistic relay selection. *IEEE Trans. Veh. Technol.* **2017**, *66*, 3930–3941. [CrossRef]
21. Wyner, A.D. The wire-tap channel. *Bell Syst. Tech. J.* **1975**, *54*, 1355–1387. [CrossRef]
22. Gopala, P.K.; Lai, L.; Gamal, H.E. On the secrecy capacity of fading channels. *IEEE Trans. Inf. Theory* **2008**, *10*, 4687–4698. [CrossRef]

23. Shafiee, S.; Liu, N.; Ulukus, S. Towards the secrecy capacity of the Gaussian MIMO wire-tap channel: The 2-2-1 channel. *IEEE Trans. Inf. Theory* **2009**, *55*, 4033–4039. [CrossRef]
24. Helena, R.P.; Jordi, H.J. Computational and energy costs of cryptographic algorithms on handheld devices. *Future Internet* **2011**, *3*, 31–48.
25. Jia, S.; Lo, M.-C.; Zhang, L.; Ozolins, O.; Udalcovs, A.; Kong, D.; Pang, X.; Yu, X.; Xiao, S.; Popov, S.; et al. Integrated dual-DFB laser for 408 GHz carrier generation enabling 131 Gbit/s wireless transmission over 10.7 meters. In Proceedings of the Optical Fiber Communication Conference, San Diego, CA, USA, 7 March 2019.
26. Guerboukha, H.; Shrestha, R.; Neronha, J.; Ryan, O.; Hornbuckle, M.; Fang, Z.; Mittleman, D.M. Efficient leaky-wave antennas at terahertz frequencies generating highly directional beams. *Appl. Phys. Lett.* **2020**, *117*, 261103. [CrossRef]
27. Qiao, J.; Alouini, M. Secure Transmission for Intelligent Reflecting Surface-Assisted mmWave and Terahertz Systems. *IEEE Wirel. Commun. Lett.* **2020**, *9*, 1743–1747. [CrossRef]
28. Qiao, J.; Zhang, C.; Dong, A.; Bian, J.; Alouini, M. Securing Intelligent Reflecting Surface Assisted Terahertz Systems. *IEEE Trans. Veh. Technol.* **2022**, *accepted*. [CrossRef]
29. Steinmetzer, D.; Chen, J.; Classen, J.; Knightly, E.; Hollick, M. Eavesdropping with periscopes: Experimental security analysis of highly directional millimeter waves. In Proceedings of the IEEE Conference on Communications and Network Security (CNS), Florence, Italy, 28–30 September 2015.
30. Ma, J.; Shrestha, R.; Adelberg, J.; Yeh, C.Y.; Hossain, Z.; Knightly, E.; Jornet, J.M.; Mittleman, D.M. Security and eavesdropping in terahertz wireless links. *Nature* **2018**, *563*, 89–93. [CrossRef]
31. Ju, Y.; Wang, H.W.; Zheng, T.X.; Yin, Q.; Lee, M.H. Safeguarding millimeter wave communications against randomly located eavesdroppers. *IEEE Trans. Wirel. Commun.* **2018**, *17*, 2675–2689. [CrossRef]
32. Wu, Y.; Kokkoniemi, J.; Han, C.; Juntti, M. Interference and coverage analysis for terahertz networks with indoor blockage effects and line-of-sight access point association. *IEEE Trans. Wirel. Commun.* **2020**, *20*, 1472–1486. [CrossRef]
33. Pinto, P.C.; Barros, J.; Win, M.Z. Wireless physical-layer security: The case of colluding eavesdroppers. In Proceedings of the IEEE International Symposium on Information Theory, Seoul, Korea, 28 June–3 July 2009.
34. Zhou, X.; Ganti, R.K.; Andrews, J.G. Secure wireless network connectivity with multi-antenna transmission. *IEEE Trans. Wirel. Commun.* **2011**, *10*, 425–430. [CrossRef]
35. Zhang, X.; Zhou, X.; McKay, M.R. Enhancing secrecy with multi-antenna transmission in wireless ad hoc networks. *IEEE Trans. Inf. Forensics Secur.* **2013**, *11*, 1802–1814. [CrossRef]
36. Xu, Q.; Ren, P.; Song, H.; Du, Q. Security enhancement for IoT communications exposed to eavesdroppers with uncertain locations. *IEEE Access* **2016**, *4*, 2840–2853. [CrossRef]
37. Zhou, X.; McKay, M.R. Secure transmission with artificial noise over fading channels: Achievable rate and optimal power allocation. *IEEE Trans. Veh. Technol.* **2010**, *59*, 3831–3842. [CrossRef]
38. Wang, H.M.; Zheng, T.; Xia, X.G. Secure MISO wiretap channels with multiantenna passive eavesdropper: Artificial noise vs. artificial fast fading. *IEEE Trans. Wirel. Commun.* **2015**, *14*, 94–106. [CrossRef]
39. Zhang, X.; McKay, M.R.; Zhou, X.; Heath, R.W. Artificial-noise-aided secure multi-antenna transmission with limited feedback. *IEEE Trans. Wirel. Commun.* **2015**, *14*, 2742–2754. [CrossRef]
40. Papasotiriou, E.N.; Boulogeorgos, A.A.A.; Alexiou, A. Performance analysis of THz wireless systems in the presence of antenna misalignment and phase noise. *IEEE Commun. Lett.* **2020**, *24*, 1211–1215. [CrossRef]
41. Ekti, A.R.; Boyaci, A.; Alparslan, A.; Ünal, İ.; Yarkan, S.; Görçin, A.; Arslan, H.; Uysal, M. Statistical modeling of propagation channels for terahertz band. In Proceedings of the IEEE Conference on Standards for Communications and Networking (CSCN), Helsinki, Finland, 18–20 September 2017.
42. Sadiku, M.N. *Numerical Techniques in Electromagnetics*, 2nd ed.; CRC Press: Boca Raton, FL, USA, 2000.
43. Stoyan, D.; Kendall, W.; Mecke, J. *Stochastic Geometry and Its Applications*, 2nd ed.; John Wiley and Sons: Hoboken, NJ, USA, 1996.
44. Johnson, W. The curious history of Faà di Bruno's formula. *Am. Math. Mon.* **2002**, *109*, 217–234.
45. Shrestha, R.; Guerboukha, H.; Fang, Z.; Knightly, E.; Mittleman, D.M. Jamming a terahertz wireless link. *Nat. Commun.* **2022**, *13*, 3045. [CrossRef]
46. Boulogeorgos, A.A.A.; Jornet, J.; Alexiou, A. Directional terahertz communication systems for 6G: Fact check: A quantitative look. *IEEE Veh. Technol. Mag.* **2021**, *16*, 68–77. [CrossRef]

MDPI
St. Alban-Anlage 66
4052 Basel
Switzerland
Tel. +41 61 683 77 34
Fax +41 61 302 89 18
www.mdpi.com

Micromachines Editorial Office
E-mail: micromachines@mdpi.com
www.mdpi.com/journal/micromachines

www.ingramcontent.com/pod-product-compliance
Lightning Source LLC
LaVergne TN
LVHW070647100526
838202LV00013B/905